Analytical and Numerical Methods
for Volterra Equations

SIAM Studies in Applied Mathematics

JOHN A. NOHEL, Managing Editor

This series of monographs focuses on mathematics and its applications to problems of current concern to industry, government, and society. These monographs will be of interest to applied mathematicians, numerical analysts, statisticians, engineers, and scientists who have an active need to learn useful methodology for problem solving.

The first six titles in this series are: *Lie-Bäcklund Transformations in Applications*, by Robert L. Anderson and Nail H. Ibragimov; *Methods and Applications of Interval Analysis*, by Ramon E. Moore; *Ill-Posed Problems for Integrodifferential Equations in Mechanics and Electromagnetic Theory*, by Frederick Bloom; *Solitons and the Inverse Scattering Transform*, by Mark J. Ablowitz and Harvey Segur; *Fourier Analysis of Numerical Approximations of Hyperbolic Equations*, by Robert Vichnevetsky and John B. Bowles; and *Numerical Solution of Elliptic Problems*, by Garrett Birkhoff and Robert E. Lynch.

Peter Linz

Analytical and Numerical Methods for Volterra Equations

siam *Philadelphia/1985*

Copyright © 1985 by Society for Industrial and Applied Mathematics. All rights reserved.

Library of Congress Catalog Card Number: 84-51968
ISBN: 0-89871-198-3

to Susan

Contents

Preface xi

Part A: Theory

Chapter 1. Introduction 3

   1.1. Classification of Volterra equations 3
   1.2. Connection between Volterra equations and initial value problems 7
   Notes on Chapter 1 10

Chapter 2. Some Applications of Volterra Equations 13

   2.1. History-dependent problems 14
   2.2. Applications in systems theory 15
   2.3. Problems in heat conduction and diffusion 19
   2.4. Some problems in experimental inference 23
   Notes on Chapter 2 26

Chapter 3. Linear Volterra Equations of the Second Kind 29

   3.1. Existence and uniqueness of the solution 29
   3.2. The resolvent kernel 35
   3.3. Some qualitative properties of the solution 38
   3.4. Systems of equations 46
   3.5. Equations with unbounded kernels 47
   3.6. Integrodifferential equations 49
   Notes on Chapter 3 50

Chapter 4. Nonlinear Equations of the Second Kind 51

   4.1. Successive approximations for Lipschitz continuous kernels . 51
   4.2. Existence and uniqueness for more general kernels 54

4.3. Properties of the solution. 58
4.4. Unbounded kernels and systems of equations 62
4.5. The resolvent equation 63
Notes on Chapter 4 65

Chapter 5. Equations of the First Kind 67

5.1. Equations with smooth kernels 67
5.2. Abel equations 71
Notes on Chapter 5 76

Chapter 6. Convolution Equations 77

6.1. Some simple kernels 77
6.2. Laplace transforms 84
6.3. Solution methods using Laplace transforms 86
6.4. The asymptotic behavior of the solution for some special equations 89
Notes on Chapter 6 92

Part B: Numerical Methods

Chapter 7. The Numerical Solution of Equations of the Second Kind . 95

7.1. A simple numerical procedure 96
7.2. Methods based on more accurate numerical integration . . . 97
7.3. Error analysis: convergence of the approximate solution . . . 100
7.4. Error estimates and numerical stability 103
7.5. Another view of stability 110
7.6. Block-by-block methods 114
7.7. Some numerical examples 118
7.8. Explicit Runge–Kutta methods 122
7.9. A summary of related ideas and methods 124
Notes on Chapter 7 127

Chapter 8. Product Integration Methods for Equations of the Second Kind 129

8.1. Product integration 130
8.2. A simple method for a specific example 132
8.3. A method based on the product trapezoidal rule 135
8.4. A block-by-block method based on quadratic interpolation . 136
8.5. A convergence proof for product integration methods 138
Notes on Chapter 8 141

Chapter 9. Equations of the First Kind with Differentiable Kernels — 143

9.1. Application of simple integration rules 144
9.2. Error analysis for simple approximation methods 145
9.3. Difficulties with higher order methods 151
9.4. Block-by-block methods 154
9.5. Use of the differentiated form 158
9.6. Nonlinear equations 160
9.7. Some practical considerations 161
Notes on Chapter 9 163

Chapter 10. Equations of the Abel Type — 165

10.1. Solving a simple Abel equation 166
10.2. The midpoint and trapezoidal methods for general Abel equations 166
10.3. Block-by-block methods 169
10.4. Some remarks on error analysis 171
10.5. Solving Abel equations in the presence of experimental errors 174
Notes on Chapter 10 175

Chapter 11. Integrodifferential Equations — 177

11.1. A simple numerical method 178
11.2. Linear multistep methods 182
11.3. Block-by-block methods 185
11.4. Numerical stability 186
11.5. Other types of integrodifferential and functional equations . 188
Notes on Chapter 11 189

Chapter 12. Some Computer Programs — 191

12.1. The trapezoidal method for systems of the second kind . . 192
12.2. The product trapezoidal method for a system of the second kind 195
12.3. The midpoint method for systems of the first kind 197
12.4. The product midpoint method for a system of the first kind. 198
Notes on Chapter 12 199

Chapter 13. Case Studies — 201

13.1. Estimating errors in the approximation 201
13.2. An example from polymer rheology 204

13.3. Solving an equation of the first kind in the presence of large
data errors 207
Notes on Chapter 13. 211

References 213

Supplementary Bibliography 223

Index 225

Preface

Starting with the work of Abel in the 1820's, analysts have had a continuing interest in integral equations. The names of many modern mathematicians, notably Cauchy, Fredholm, Hilbert, Volterra, and others, are associated with this topic. There are basically two reasons for this interest. In some cases, as in the work of Abel on tautochrone curves, integral equations are the natural mathematical model for representing a physically interesting situation. The second, and perhaps more common reason, is that integral operators, transforms, and equations, are convenient tools for studying differential equations. Consequently, integral equation techniques are well known to classical analysts and many elegant and powerful results were developed by them.

Amongst applied mathematicians, engineers, and numerical analysts a working knowledge of integral equations is less common. In mathematical modeling, the traditional emphasis has been on differential equations. Analytical and numerical techniques for differential equations are widely known by many who are essentially unfamiliar with integral equation techniques. A brief glance at standard numerical analysis texts will confirm this; numerical methods for integral equations are rarely mentioned. Recently, however, this situation has begun to change. As mathematical models become more realistic, integral equations may become unavoidable. In some cases, as in problems with delayed action, differential equations can no longer represent the physically essential properties, and integral terms have to be introduced. In other cases, it is known that integral equations provide a convenient and practically useful alternative to differential equations. This has led to the currently active field of boundary integral equation methods which has attracted many engineers. Numerical analysts have followed suit, and the study of numerical methods for integral equations has become a topic of considerable interest.

Our aim in this book is to present one aspect of this recent activity in integral equations, namely, methods for the solution of Volterra equations.

Arising primarily in connection with history-dependent problems, such equations have been studied since the work of Volterra on population dynamics. Since there are few known analytical methods leading to closed-form solutions, our emphasis will be on numerical techniques. The second part of the book is devoted entirely to numerical methods. Still, some analytical techniques for studying the properties of the solution are known. These are important not only for gaining insight into the qualitative behavior of the solution, but are also essential in the design of effective numerical methods. The major points of these analytical methods are presented in the first part of the book.

Much of the work on the numerical solution of Volterra equations was carried out in the twenty year period between 1960 and 1980. Before 1960, only sporadic attempts were made to develop numerical algorithms. A systematic attack on the whole problem started with the work of Pouzet in about 1960. Although later research greatly extended and improved Pouzet's work, its merit as a catalyst is considerable. By 1980, most of the major principles were well understood and the study had attained some maturity. The list of references at the end of the book shows this, with most entries dating from this period. Currently, there is still a considerable interest in this field, although now special problems rather than a general theory have become the focus of attention. Extensions to partial integrodifferential equations or other functional equations, as well as questions of numerical stability, are the concern of most of the very recent papers. The supplementary bibliography, dealing with work published after 1982, reflects this current interest.

The audience for which this book is intended is a practical one with an immediate need to solve real-world problems. With this in mind, I have chosen the simplest possible setting for the discussion, the space of real functions of real variables. No attempt has been made to extend the results to more general function spaces, although this is certainly possible. As a result, a good grasp of calculus is sufficient for understanding almost all of the given material. I have used examples (sometimes very simple) where I felt that the general discussion could be illuminated by a specific instance. Also, I have included a number of exercises with a two-fold purpose. First, they can serve as a test of the reader's understanding. They are usually simple enough to give no trouble to those who have firmly grasped the preceding discussion. The second purpose of the exercises is to allow me to present extensions without having to give proofs or other lengthy discussions. In many cases (here as well as in other areas), generalizations of known theorems are often technically tedious, but conceptually straightforward. I prefer to leave this to the interested reader.

My interest and understanding of this field developed over a long period of time, starting with my Ph.D. dissertation in 1968. During the intervening years I have talked about this topic with most of the researchers active in the field. My thanks go to all of them for their contributions and many stimulating discussions. I am also indebted to the reviewers of the manuscript, amongst them C. T. H. Baker, J. M. Bownds, and A. Gerasoulis, for comments which were valuable in the preparations of the final version.

Finally, I would like to acknowledge a debt which goes well beyond the usual one incurred by authors. It is almost certain that without the continued interest and help of Ben Noble this book would never have been written. Ben introduced me to Volterra equations when, as a new graduate student in 1965, I was searching for an appropriate thesis topic. His advice and direction were largely responsible for my initial contributions to the field. The writing of this book was started by a suggestion from him and was originally conceived as a joint effort. When Ben had to withdraw from the project later, he graciously consented to let me use the already significant amount of work he had done. While in the end I wrote all of the book, its underlying structure as well some of the specific approaches were very much influenced by this early work. Thus, although his name does not appear on the title page of the book, the contribution of Ben Noble to this work is considerable.

PETER LINZ
Davis, California
July 1984

PART A: THEORY

Chapter 1

Introduction

An integral equation is an equation in which the unknown, generally a function of one or more variables, occurs under an integral sign. This rather general definition allows for many different specific forms and in practice many distinct types arise.

In the classical theory of integral equations one distinguishes between *Fredholm* equations and *Volterra* equations. In a Fredholm equation the region of integration is fixed, whereas in a Volterra equation the region is variable, usually depending in some simple fashion on the independent variables. Some integral equations are called *singular*, although the usage of this term varies. Some authors call an equation singular if the integrals cannot be interpreted in the usual way (i.e., in the Riemann or Lebesgue sense), but must be considered as principal value integrals. Equations which have unbounded but integrable integrands are then called *weakly* singular. Others use the term singular to denote any equation in which some of the integrands are unbounded. We will not be interested in principal value equations here and will use the adjective singular in its second sense.

The distinction between Fredholm and Volterra equations is analogous to the distinction between boundary and initial value problems in ordinary differential equations. Volterra equations can be considered a generalization of initial value problems. In practice, Volterra equations frequently occur in connection with time-dependent or *evolutionary* systems. However, this is not always the case as is shown by examples in later chapters.

1.1. Classification of Volterra equations. An equation of the form

$$f(t) - \int_a^t K(t, s, f(s))\, ds = g(t), \qquad a \leq t \leq T, \tag{1.1}$$

is a Volterra equation of the *second kind*. Here the unknown is $f(s)$. The right-hand side $g(t)$ and the *kernel* $K(t, s, u)$ are assumed to be known. Equation (1.1) is one of several forms in which a Volterra equation can be written. More generally, one might consider the form

$$F\left(f(t), t, \int_a^t K(t, s, f(s))\, ds, g(t)\right) = 0, \qquad (1.2)$$

but we will limit our attention to the more common form (1.1).

For our purposes we assume that T is finite. In many practical applications, the behavior of the solution on the whole real axis is of interest. In this situation the limiting behavior of the solution is usually found from its behavior for large, but finite T. In numerical computations it is necessary in any case to use a finite T.

For notational simplicity we can, without loss of generality, choose the range of the independent variable so that the lower limit is zero and consider only the equation

$$f(t) - \int_0^t K(t, s, f(s))\, ds = g(t), \qquad 0 \leq t \leq T. \qquad (1.3)$$

In our subsequent discussion, whenever the domain of the equation is unspecified, we will assume it to be $0 \leq t \leq T < \infty$.

Of special interest is the linear case in which

$$K(t, s, f(s)) = k(t, s) f(s). \qquad (1.4)$$

Linearity somewhat simplifies the treatment of the equation, although when the nonlinearity is suitably restricted it introduces few essential complications.

There are many applications where the kernel of the equation is unbounded, that is, the equation is (in our terminology) singular. Where possible, we will write the kernels of such equations as

$$K(t, s, f(s)) = p(t, s) H(t, s, f(s)),$$

where $p(t, s)$ represents the singular part, that is, it is chosen so that $H(t, s, f(s))$ is bounded.

A fundamentally different kind of equation is the Volterra equation of the *first kind*

$$\int_0^t k(t, s, f(s))\, ds = g(t). \qquad (1.5)$$

Although formally one can often reduce such an equation to one of the second kind (e.g., by differentiation), we will see in subsequent discussions that equations of the first kind present some serious practical difficulties.

Historically, one of the earliest integral equations to be studied was *Abel's equation*

$$\int_0^t \frac{f(s)}{\sqrt{t-s}}\, ds = g(t), \qquad (1.6)$$

which is an example of a singular equation of the first kind. Nowadays it is fairly common practice to call the equation

$$\int_0^t p(t,s)h(t,s)f(s)\,ds = g(t), \tag{1.7}$$

with $h(t,s)$ bounded and $p(t,s)$ unbounded (but restricted to guarantee existence and uniqueness of the solution) a *generalized Abel equation*.

Formally, one can immediately extend the classification to systems of equations by interpreting f, K and g as vectors. Thus (1.3) becomes

$$\mathbf{f}(t) - \int_0^t \mathbf{K}(t,s,\mathbf{f}(s))\,ds = \mathbf{g}(s), \tag{1.8}$$

where

$$\mathbf{f}(s) = \begin{pmatrix} f_1(s) \\ f_2(s) \\ \vdots \\ f_m(s) \end{pmatrix}, \tag{1.9}$$

and

$$\mathbf{K}(t,s,\mathbf{f}(s)) = \begin{pmatrix} K_1(t,s,f_1(s),f_2(s),\ldots,f_m(s)) \\ K_2(t,s,f_1(s),f_2(s),\ldots,f_m(s)) \\ \vdots \\ K_m(t,s,f_1(s),f_2(s),\ldots,f_m(s)) \end{pmatrix}. \tag{1.10}$$

Equation (1.8) is then a system of the second kind.

Volterra *integrodifferential equations* involve derivatives of the unknown as well as integral terms. The presence of both derivatives and integrals allows for a profusion of different forms, but there does not exist any commonly used convention for classifying them. Fortunately, most of the equations arising in practice have a fairly simple form and can usually be reduced to integral equations. We will use this observation in later chapters; for the moment, we only illustrate it with some examples.

Example 1.1. Consider the integrodifferential equation

$$f'(t) - \int_0^t K(t,s,f(s))\,ds = g(t), \tag{1.11}$$

$$f(0) = f_0.$$

Introducing the new function

$$f'(t) = z(t),$$

we get

$$z(t) - \int_0^t K(t, s, f(s)) \, ds = g(t), \quad (1.12)$$

$$f(t) - \int_0^t z(s) \, ds = f_0, \quad (1.13)$$

so that (1.11) is equivalent to a system of the second kind.

Example 1.2. For certain types of linear integrodifferential equations, the reduction can be made directly by integration. Consider, for instance, the linear equation

$$f'(t) - \int_0^t k(t, s) f(s) \, ds = g(t), \quad (1.14)$$

with $f(0) = f_0$. Integrating this we get

$$f(t) - \int_0^t \int_0^\tau k(\tau, s) f(s) \, ds \, d\tau = G(t), \quad (1.15)$$

where

$$G(t) = f_0 + \int_0^t g(\tau) \, d\tau. \quad (1.16)$$

We can then interchange the order of integration in (1.15), using elementary calculus arguments (Fig. 1.1), to obtain

$$f(t) - \int_0^t M(t, s) f(s) \, ds = G(t), \quad (1.17)$$

where

$$M(t, s) = \int_s^t k(\tau, s) \, d\tau. \quad (1.18)$$

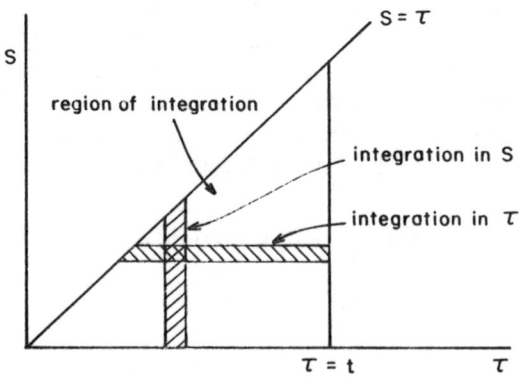

FIG. 1.1. *Interchanging the order of integration.*

INTRODUCTION

Such arguments can also be used to reduce integrodifferential equations of higher order to integral equations.

Exercise 1.1. Reduce the equation

$$f''(t) + b(t)f(t) + \int_0^t k(t, s)f(s)\, ds = g(t),$$
$$f(0) = \alpha, \qquad (1.19)$$
$$f'(0) = \beta$$

to a system of Volterra equations.

Exercise 1.2. Use direct integration to reduce (1.19) to a single Volterra equation.

1.2. Connection between Volterra equations and initial value problems. Integral equations are used extensively in the study of the properties of differential equations. The most elementary observation is that the differential equation

$$y'(t) = F(t, y(t)), \qquad t \geq 0,$$
$$y(0) = y_0 \qquad (1.20)$$

can be converted by integration into the Volterra equation

$$y(t) = y_0 + \int_0^t F(s, y(s))\, ds. \qquad (1.21)$$

This is often the starting point for the exploration of the qualitative properties of the solution of (1.20), for example, its existence and uniqueness.

Our concern here is not the conversion of differential to integral equations, but the converse. When the kernel of the integral equation has certain special properties, it is possible to find an equivalent system of ordinary differential equations. In extremely simple examples, this gives a way for finding closed form solutions to Volterra equations.

As the most elementary case consider

$$f(t) + \int_0^t e^{\alpha(t-s)} f(s)\, ds = g(t). \qquad (1.22)$$

Differentiating both sides, we get

$$f'(t) + f(t) + \alpha \int_0^t e^{\alpha(t-s)} f(s)\, ds = g'(t).$$

Substituting for the integral term from (1.22) gives

$$f'(t) + (1 - \alpha)f(t) = g'(t) - \alpha g(t).$$

From (1.22), the initial value for this differential equation is
$$f(0) = g(0).$$
This example is a special case of a linear equation with kernel
$$k(t, s) = -P(t)Q(s), \tag{1.23}$$
with $P(t) \neq 0$ in $[0, T]$. If we substitute this form into (1.3), we have the equation
$$f(t) + \int_0^t P(t)Q(s)f(s)\,ds = g(t). \tag{1.24}$$
If we divide by $P(t)$ and introduce the variable $u(t) = f(t)/P(t)$, the equation becomes
$$u(t) + \int_0^t P(s)Q(s)u(s)\,ds = \frac{g(t)}{P(t)}. \tag{1.25}$$
Differentiation then yields the equation
$$u'(t) + P(t)Q(t)u(t) = \frac{d}{dt}\left[\frac{g(t)}{P(t)}\right], \tag{1.26}$$
with
$$u(0) = \frac{g(0)}{P(0)}.$$
The elementary integrating factor method immediately gives the solution of (1.26) as
$$u(t) = \frac{1}{v(t)}\left\{\int_0^t v(s)\frac{d}{ds}\left[\frac{g(s)}{P(s)}\right]ds + \frac{g(0)}{P(0)}\right\},$$
with
$$v(t) = \exp\left\{\int_0^t P(s)Q(s)\,ds\right\}.$$
This can further be simplified by integration by parts to
$$u(t) = \frac{g(t)}{P(t)} - \frac{1}{v(t)}\int_0^t v(s)Q(s)g(s)\,ds,$$
and the solution
$$f(t) = g(t) - \frac{P(t)}{v(t)}\int_0^t v(s)Q(s)g(s)\,ds. \tag{1.27}$$
This result can be generalized to kernels of the form
$$k(t, s) = -\sum_{i=1}^n P_i(t)Q_i(s), \tag{1.28}$$
which are usually referred to as *degenerate* or *finite-rank* kernels.

THEOREM 1.1. *Let*

$$k(t, s) = -\sum_{i=1}^{n} P_i(t) Q_i(s).$$

Assume that $P_i(t)$, $Q_i(t)$ and a given function $g(t)$ are continuous in $[0, T]$. Then the linear equation

$$f(t) - \int_0^t k(t, s) f(s) \, ds = g(t) \tag{1.29}$$

has a solution

$$f(t) = g(t) - \sum_{i=1}^{n} P_i(t) y_i(t), \tag{1.30}$$

where the $y_i(t)$ are the solution of the system

$$y_i'(t) = Q_i(t) \left\{ g(t) - \sum_{j=1}^{n} P_j(t) y_j(t) \right\}, \quad i = 1, 2, \ldots, n, \tag{1.31}$$

$$y_i(0) = 0. \tag{1.32}$$

Proof. Let $y_i(t)$ be the solution of (1.31) subject to the initial conditions (1.32). Because of the assumptions made, the $y_i(t)$ exist and are continuous. Then, integrating (1.31), using (1.32) gives

$$y_i(t) = \int_0^t Q_i(s) \left\{ g(s) - \sum_{j=1}^{n} P_j(s) y_j(s) \right\} ds.$$

Now define $u(t)$ as

$$u(t) = g(t) - \sum_{i=1}^{n} P_i(t) y_i(t). \tag{1.33}$$

Then

$$\int_0^t k(t, s) u(s) \, ds = -\int_0^t \sum_{i=1}^{n} P_i(t) Q_i(s) \left\{ g(s) - \sum_{j=1}^{n} P_j(s) y_j(s) \right\} ds$$

$$= -\sum_{i=1}^{n} P_i(t) \int_0^t Q_i(s) \left\{ g(s) - \sum_{j=1}^{n} P_j(s) y_j(s) \right\} ds$$

$$= -\sum_{i=1}^{n} P_i(t) \int_0^t y_i'(s) \, ds$$

$$= -\sum_{i=1}^{n} P_i(t) y_i(t).$$

Therefore,

$$\int_0^t k(t, s) u(s) \, ds = u(t) - g(t),$$

showing that $u(t)$ satisfies (1.29). Since $u(t)$ has the form (1.30), the proof is completed. □

In Chapter 3 it will be shown that, under the stated assumptions, equation (1.29) has a unique solution. Therefore the solution of the system (1.31), together with (1.30), gives the unique solution of a linear Volterra equation of the second kind with a degenerate kernel.

The result of Theorem 1.1 can be extended to certain nonlinear equations, as is indicated by the following exercise.

Exercise 1.3. Show that, under the appropriate assumptions, the equation

$$f(t) - \int_0^t K(t, s, f(s))\, ds = g(t),$$

with

$$K(t, s, u) = -\sum_{i=1}^n P_i(t) Q_i(s, u),$$

has a solution

$$f(t) = g(t) - \sum_{i=1}^n P_i(t) y_i(t),$$

with $y_i(t)$ satisfying

$$y_i'(t) = Q_i\left(t, g(t) - \sum_{j=1}^n P_j(t) y_j(t)\right),$$

$$y_i(0) = 0.$$

Notes on Chapter 1. There are many texts on integral equations, but most treat Volterra equations only briefly. To some extent this is because Volterra equations can be considered a special case of Fredholm equations. For example,

$$f(t) - \int_0^t k(t, s) f(s)\, ds = g(t), \qquad 0 \le t \le 1,$$

can be written as a Fredholm equation

$$f(t) - \int_0^1 k(t, s) f(s)\, ds = g(t),$$

if we set $k(t, s) = 0$ for $s > t$. The classical Fredholm theory therefore also applies to Volterra equations, but loses much of its power because the kernel is not symmetric. A direct study of Volterra equations yields many results which cannot be obtained with the Fredholm theory.

Some books which make more than a passing reference to Volterra equations are Cochran [68], Corduneanu [70], Davis [75], Kowalewski [155], Mikhlin [187], Pogorzelski [207], Tricomi [229], Volterra [232], and

Yosida [247]. Tsalyuk [230] contains many references on analytical and numerical work, with special emphasis on the Russian literature.

The equivalence between Volterra equations with degenerate kernels and systems of differential equations is mentioned by Cochran [68], who attributes the observation to Goursat. A more detailed discussion can be found in [30]. On the whole, the reduction seems to have been ignored until it was used by Bownds and his coworkers to construct methods for the approximate solution of Volterra equations (see for example [34], [35]).

Chapter 2

Some Applications of Volterra Equations

Before setting out on a detailed investigation of Volterra equations, we briefly look at some actual applications where such equations arise. Our aim here will be to develop some understanding of the usefulness of Volterra integral equations without undue attention to detail. Consequently, the discussion will be somewhat intuitive and at times simplified, with mathematical arguments that are quite informal. Nevertheless, the examples given are representative of a variety of more complicated applications.

Volterra equations arise most naturally in certain types of time-dependent problems whose behavior at time t depends not only on the state at that time, but also on the states at previous times. The solution of the differential equation

$$y'(t) = f(t, y(t))$$

is completely determined for $t > t_0$ if $y(t_0)$ is known. This is true for any t_0. Information prior to $t = t_0$ is irrelevant to the solution after t_0. There are, however, situations where knowledge of the current state alone is not enough, and where it is necessary to know how the state $y(t_0)$ is arrived at in order to predict the future. Such models are sometimes called *history-dependent* or *systems with memory*. If the history dependence can be represented by a term

$$\int_0^t K(t, s, y(s)) \, ds,$$

then the modeling equation is of Volterra type. Some elementary examples of such systems are discussed in §§ 2.1 and 2.2.

Integral equations also find their use in applications where the more obvious model is a differential equation. There may be several advantages to reducing (when possible) a differential equation. From a theoretical point of view, integral operators are more easily dealt with than differential operators, and properties of the solution may be more readily inferred from the integral form. The simplest and best known example, already mentioned in § 1.2, is the reduction of the ordinary differential (1.20) to the integral equation (1.21). There may also be some practical advantages. In some cases the integral equation reduces the dimensionality. For example, certain partial differential equations in two variables can be shown to be equivalent to integral equations in one variable, thereby considerably simplifying the numerical computations. Several examples of this are given in § 2.3.

Finally, integral equations arise in some situations where experimental observations yield not the variable of interest but rather some integrals thereof. To compute the actual variable then requires the solution of integral equations. Section 2.4 contains some examples of this type.

2.1. History-dependent problems. One of the best known examples of this type is the so-called *renewal equation*. Consider a component of some machine which is subject to failure as time passes. In general, the failure time is a random variable characterized by a probability density $\rho(t)$ such that, in the small interval $(t, t+\Delta t)$ the probability of failure of a component which is new at t' is

$$\rho(t-t')\,\Delta t. \tag{2.1}$$

If every component eventually fails then

$$\int_{t'}^{\infty} \rho(t-t')\,dt = 1, \tag{2.2}$$

for every t', so that $\rho(t)$ must satisfy

$$\int_{0}^{\infty} \rho(t)\,dt = 1. \tag{2.3}$$

Assume now that as soon as the component fails it is replaced by a new one. This new component will be replaced when it fails, and so on. Of practical interest is the *renewal density* $h(t)$ which measures the probability for the need of a replacement. It is defined so that the probability that a renewal has to be made in the interval $(t, t+\Delta t)$ is given by $h(t)\,\Delta t$. The probability for a needed replacement is the sum of (a) the probability that the first failure occurs in $(t, t+\Delta t)$, and (b) the probability that a renewal was made at time t', followed by another failure after $t-t'$ time units. If all contributions are added and the limit $\Delta t \to 0$ taken, we get the equation

$$h(t) = \rho(t) + \int_{0}^{t} h(t')\rho(t-t')\,dt'. \tag{2.4}$$

This is the renewal equation. It is a Volterra equation of the second kind of a particularly simple form.

The study of Volterra equations originated with the work of V. Volterra on population dynamics. The classical, elementary *population equation* is

$$\frac{dN(t)}{dt} = \alpha N(t), \qquad t \geq 0, \tag{2.5}$$

$$N(0) = N_0.$$

Here $N(t)$ stands for the number of individuals of a population alive at time t. Equation (2.5) states the assumption that the rate of change of the population, that is, the number of births and deaths at time t, depends only on the number of individuals alive at time t. The constant α represents the difference between the birth and death rates; if $\alpha > 0$ the population grows exponentially, if $\alpha < 0$ it will die out at an exponential rate.

The simple assumptions leading to equation (2.5) are rarely realistic and more complicated models have to be constructed to account for observed phenomena. For example, the environment in which the population exists may change due to factors such as pollution or exhaustion of the food supply. In such a case, the growth factor α may change with the state of the environment. The state of the environment at time t in turn depends on the past history of the population, since this population contributes to the change in the environment. Thus, instead of a constant α, it becomes plausible to use a variable growth factor incorporating a history-dependent term, for example,

$$\alpha(t) = \alpha_0 - \int_0^t k(t-s)N(s)\,ds. \tag{2.6}$$

Equation (2.5) then becomes the Volterra integrodifferential equation

$$\frac{dN(t)}{dt} = N(t)\left\{\alpha_0 - \int_0^t k(t-s)N(s)\,ds\right\}. \tag{2.7}$$

The equation actually used by Volterra in the original study was not (2.7), but an equation of the form

$$\frac{dN(t)}{dt} = N(t)\left\{\alpha_0 - \alpha_1 N(t) - \int_0^t k(t-s)N(s)\,ds\right\}. \tag{2.8}$$

The additional term $-\alpha_1 N^2(t)$ was introduced to account for the competition between individuals in the population. This term, which has a negative coefficient, tends to inhibit the growth of the population.

2.2. Applications in systems theory. The term *system* is used rather generally to denote a physical or otherwise observable entity operating according to certain principles. Basically, we think of a system as a process which transforms a given input into a proper response or output. Some

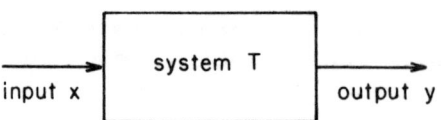

FIG. 2.1. *Schema of a simple system.*

simple examples are mechanical systems, electric circuits, and computer systems. The concept is quite broad and includes a great variety of complex phenomena. From the mathematical point of view a system can be characterized by three elements: a linear space X of all possible inputs, a linear space Y of outputs, and a transformation T which maps elements of X into Y. In other words, a system can be described formally by the equation

$$Tx = y. \tag{2.9}$$

Schematically, such a system is depicted in Fig. 2.1. This description is clearly so general and all-inclusive that little of interest can be done with it. To proceed we must be more specific.

First of all, we choose as X and Y the space of functions of a single real variable t. We interpret t as "time"; in many applications, particularly those for which Volterra equations are important, the independent variable is indeed the elapsed time measured from some chosen reference time $t = 0$. Systems can be classified as *linear* or *nonlinear*; a linear system is one in which the superposition principle holds. Linear systems are considerably more tractable than nonlinear ones. We shall therefore consider the system (2.9), where for all elements of X, T satisfies

$$T(x_1 + x_2) = T(x_1) + T(x_2), \tag{2.10}$$

$$T(\alpha x) = \alpha T(x), \tag{2.11}$$

which is essentially the superposition principle. Although the setting is still quite general, we can now show that for many such systems T can be represented as an integral operator of the Volterra type. To see this, we introduce the *unit-impulse function*

$$\sigma_\Delta(t, t_i) = \begin{cases} 1, & t_i - \dfrac{\Delta}{2} \leq t \leq t_i + \dfrac{\Delta}{2}, \\ 0 & \text{otherwise.} \end{cases} \tag{2.12}$$

This impulse function has value one during a small time period centered at t_i; everywhere else it is zero. If it is put through the system, the response generated by it must satisfy certain conditions. First, because of the linearity, the response must be proportional to Δ and we can write

$$T\sigma_\Delta(t, t_i) = \Delta m(t, t_i).$$

The function $m(t, t_i)$ is the *unit-response function*, which characterizes the response of the system at time t to a unit impulse applied at time t_i. Furthermore, in time-dependent systems one expects that the response cannot be felt before the signal is applied. The general principle of *causality* can be expressed by requiring that

$$m(t, t_i) = 0 \quad \text{if } t \leq t_i - \Delta/2. \tag{2.13}$$

Consider now what happens when a given input $f(t)$ is applied during time $0 \leq t \leq A$. If we pick N large enough and let $\Delta = A/N$, $t_i = (i - \tfrac{1}{2}) \Delta$, then $f(t)$ can be closely approximated by

$$f(t) \simeq \sum_{i=1}^{N} f(t_i) \sigma_\Delta(t, t_i),$$

so that

$$Tf(t) \simeq T \sum_{i=1}^{N} f(t_i) \sigma_\Delta(t, t_i) \simeq \Delta \sum_{i=1}^{N} m(t, t_i) f(t_i).$$

If we now take the limit $\Delta \to 0$, $N \to \infty$ then, provided the limits exist, we see that

$$Tf(t) = \int_0^A m(t, \tau) f(\tau) \, d\tau = \int_0^t m(t, \tau) f(\tau) \, d\tau,$$

where the second step uses the causality condition (2.13). These somewhat intuitive considerations then make it plausible that a general linear causal system can be described by the Volterra integral equation

$$\int_0^t m(t, \tau) f(\tau) \, d\tau = g(t), \tag{2.14}$$

where $f(t)$ is the input to the system and $g(t)$ the resulting output.

A further simplification can be achieved by considering systems whose underlying structure does not change with time. In such a system we expect that the response at time t to a signal applied at time τ depends only on the elapsed time $t - \tau$ between the input signal and the output response. In this case $m(t, \tau) = m(t - \tau)$. This is called a linear *time-invariant* causal system and is described by an equation of the form

$$\int_0^t m(t - \tau) f(\tau) \, d\tau = g(t). \tag{2.15}$$

An equation such as this, for which the kernel depends only on the difference $t - \tau$ is called a *convolution* equation. The treatment of convolution equations is often much easier than that of the general case.

Starting with these quite general observations, we can now identify some subclasses of problems in system theory. In system *identification* the properties of the system are assumed to be unknown and are to be inferred from observations, in particular, a known correspondence between an input $f(t)$ and output $g(t)$. The time-invariant case is particularly simple. If in (2.15) the transformation of variables $\sigma = t - \tau$ is made, then the equation becomes

$$\int_0^t f(t-\sigma)m(\sigma)\,d\sigma = g(t), \tag{2.16}$$

which is a Volterra equation of the first kind for the unknown $m(\sigma)$.

A type of a system of considerable practical interest is generated by connecting two simpler systems. *Feedback* systems are constructed by "feeding back" the output, suitably transformed, to the system. A schematic representation of a simple feedback system is given in Fig. 2.2.

If we restrict ourselves to linear, time-invariant systems, then the situation depicted in Fig. 2.2 can be described by the set of equations

$$\int_0^t m_1(t-s)\{f(s) + u(s)\}\,ds = g(t), \tag{2.17}$$

$$\int_0^t m_2(t-s)g(s)\,ds = u(t), \tag{2.18}$$

where $m_1(t)$ and $m_2(t)$ are the respective unit-response functions for T_1 and T_2. If the main interest is in the response $g(t)$, the unknown $u(t)$ can be eliminated by substituting (2.17) into (2.18), giving

$$\int_0^t m_1(t-s)\left\{f(s) + \int_0^s m_2(s-\tau)g(\tau)\,d\tau\right\}ds = g(t). \tag{2.19}$$

Interchanging orders of integration, we obtain

$$g(t) - \int_0^t k(t,\tau)g(\tau)\,d\tau = \int_0^t m_1(t-s)f(s)\,ds, \tag{2.20}$$

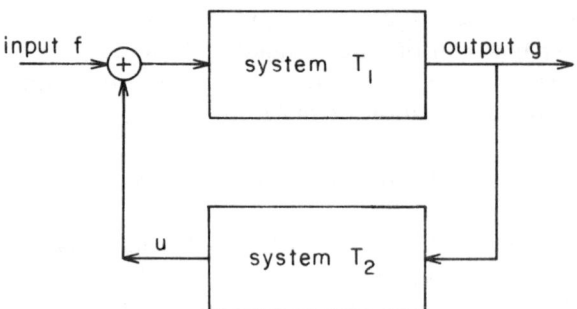

FIG. 2.2. *Schema of a feedback system.*

where

$$k(t, \tau) = \int_\tau^t m_1(t-s)m_2(s-\tau)\, ds. \tag{2.21}$$

If the properties of the two simpler systems, that is, their unit-response functions $m_1(t)$ and $m_2(t)$, and the input $f(s)$ are known, then (2.20) is a Volterra equation of the second kind for $g(t)$, the response of the feedback system. In fact, $k(t, \tau)$ is a function only of $t-\tau$, so that the equation is of convolution type.

Exercise 2.1. Show that $k(t, \tau)$ defined by (2.21) is a function of $t-\tau$ only.

2.3. Problems in heat conduction and diffusion. The recasting of differential equations in terms of integral equations can be done not only in simple cases such as equation (1.20), but also for certain partial differential equations. This is particularly common in equations of parabolic type encountered in various heat conduction and diffusion problems.

The method of reduction utilizes Fourier transforms (although sometimes Laplace transforms can be used as well). In our discussion we will use the following notation for the Fourier sine integral, the Fourier cosine integral, and the full Fourier integral:

$$\hat{f}_s(\omega) = \sqrt{\frac{2}{\pi}} \int_0^\infty f(x) \sin \omega x \, dx, \tag{2.22}$$

$$\hat{f}_c(\omega) = \sqrt{\frac{2}{\pi}} \int_0^\infty f(x) \cos \omega x \, dx, \tag{2.23}$$

$$\hat{f}(\omega) = \frac{1}{\sqrt{2\pi}} \int_{-\infty}^\infty f(x) e^{i\omega x} \, dx. \tag{2.24}$$

The corresponding inversion formulas then are

$$f(x) = \sqrt{\frac{2}{\pi}} \int_0^\infty \hat{f}_s(\omega) \sin \omega x \, d\omega \tag{2.25}$$

$$= \sqrt{\frac{2}{\pi}} \int_0^\infty \hat{f}_c(\omega) \cos \omega x \, d\omega \tag{2.26}$$

$$= \frac{1}{\sqrt{2\pi}} \int_{-\infty}^\infty \hat{f}(\omega) e^{-i\omega x} \, d\omega. \tag{2.27}$$

Consider now the simple heat conduction equation

$$\frac{\partial^2 u}{\partial x^2} = \frac{\partial u}{\partial t}, \quad 0 \le x < \infty, \quad t > 0, \tag{2.28}$$

subject to the conditions

$$u(x, 0) = 0, \qquad 0 \leq x < \infty, \tag{2.29}$$

$$\frac{\partial u(0, t)}{\partial x} = -g(t), \qquad t > 0, \tag{2.30}$$

with the further assumption that, for all t,

$$\lim_{x \to \infty} u(x, t) = 0,$$

$$\lim_{x \to \infty} \frac{\partial u(x, t)}{\partial x} = 0.$$

If we apply the Fourier cosine transformation to equation (2.28), we obtain

$$\int_0^\infty \frac{\partial^2 u(x, t)}{\partial x^2} \cos \omega x\, dx = \int_0^\infty \frac{\partial u(x, t)}{\partial t} \cos \omega x\, dx.$$

Integrating the left-hand side by parts twice, and using (2.30) and the prescribed conditions at infinity, yields

$$g(t) - \omega^2 \int_0^\infty u(x, t) \cos \omega x\, dx = \frac{\partial}{\partial t} \int_0^\infty u(x, t) \cos \omega x\, dx,$$

or

$$\sqrt{\frac{\pi}{2}} \frac{\partial}{\partial t} \hat{u}_c(\omega, t) = g(t) - \sqrt{\frac{\pi}{2}} \omega^2 \hat{u}_c(\omega, t). \tag{2.31}$$

Also, from (2.29)

$$\hat{u}_c(\omega, 0) = 0. \tag{2.32}$$

The differential equation (2.31) subject to the initial condition (2.32) has an elementary solution

$$\hat{u}_c(\omega, t) = \sqrt{\frac{2}{\pi}} e^{-\omega^2 t} \int_0^t g(s) e^{\omega^2 s}\, ds. \tag{2.33}$$

Applying now the inversion formula (2.26), we have

$$u(x, t) = \frac{2}{\pi} \int_0^\infty e^{-\omega^2 t} \int_0^t g(s) e^{\omega^2 s} \cos \omega x\, ds\, d\omega$$

$$= \frac{2}{\pi} \int_0^t g(s) \int_0^\infty e^{-\omega^2 (t-s)} \cos \omega x\, d\omega\, ds.$$

The inner integral can be explicitly evaluated as

$$\int_0^\infty e^{-\omega^2 (t-s)} \cos \omega x\, d\omega = \frac{\sqrt{\pi} e^{-x^2/4(t-s)}}{2\sqrt{t-s}},$$

so that

$$u(x, t) = \frac{1}{\sqrt{\pi}} \int_0^t g(s)(t-s)^{-1/2} e^{-x^2/4(t-s)} \, ds. \tag{2.34}$$

So far everything is standard and well known. The problem becomes more interesting when we complicate it. The condition (2.30) gives the gradient of u (and hence the rate of transfer across the boundary) as a fixed function of time. In a physical situation where u represents temperature or concentration this is not very realistic. Usually the gradient depends on the surface temperature or concentration, so that we must replace $g(t)$ with some function $G(u(0, t), t)$ and (2.34) becomes

$$u(x, t) = \frac{1}{\sqrt{\pi}} \int_0^t G(u(0, s), s)(t-s)^{-1/2} e^{-x^2/4(t-s)} \, ds. \tag{2.35}$$

If we now use $U(t)$ for $u(0, t)$ and put $x = 0$ in (2.35), we get a nonlinear Volterra equation of the second kind with an unbounded kernel

$$U(t) = \frac{1}{\sqrt{\pi}} \int_0^t \frac{G(U(s), s)}{\sqrt{t-s}} \, ds. \tag{2.36}$$

Thus, the behavior of the solution on the boundary is governed by an integral equation in a single variable.

Another example of this type occurs in nuclear reactor dynamics. The relation between the temperature of the reactor $T(x, t)$, and the power produced $u(t)$, can be described by the rather complicated set of integro-partial differential equations

$$\frac{du(t)}{dt} = \int_{-\infty}^{\infty} \alpha(x) T(x, t) \, dx, \tag{2.37}$$

$$\frac{\partial T(x, t)}{\partial t} = \frac{\partial^2 T(x, t)}{\partial x^2} + \eta(x) u(t), \tag{2.38}$$

for $-\infty < x < \infty$, $t > 0$. Additional conditions will be taken as

$$u(0) = 0,$$

$$T(x, 0) = f(x),$$

$$\lim_{x \to \pm \infty} T(x, t) = \lim_{x \to \pm \infty} \frac{\partial}{\partial x} T(x, t) = 0.$$

The second of these equations, (2.38), is simply a diffusion equation with an added source term due to the power generated by the reactor. Equation (2.37) expresses the power production as a function of the temperature.

Applying the full Fourier transform to (2.38), we get

$$\frac{\partial}{\partial t}\int_{-\infty}^{\infty} T(x,t)e^{i\omega x}\,dx = \int_{-\infty}^{\infty} \frac{\partial^2 T(x,t)}{\partial x^2} e^{i\omega x}\,dx + u(t)\int_{-\infty}^{\infty} \eta(x)e^{i\omega x}\,dx,$$

which, using integration by parts and the condition at infinity, becomes

$$\frac{\partial}{\partial t}\int_{-\infty}^{\infty} T(x,t)e^{i\omega x}\,dx = -\omega^2 \int_{-\infty}^{\infty} T(x,t)e^{i\omega x}\,dx + u(t)\int_{-\infty}^{\infty} \eta(x)e^{i\omega x}\,dx.$$

Therefore, the Fourier transform $\hat{T}(\omega, t)$ satisfies

$$\frac{\partial}{\partial t}\hat{T}(\omega, t) = -\omega^2 \hat{T}(\omega, t) + u(t)\hat{\eta}(\omega), \qquad (2.39)$$

with initial condition

$$\hat{T}(\omega, 0) = \hat{f}(\omega).$$

This equation has the simple solution

$$\hat{T}(\omega, t) = e^{-\omega^2 t}\left\{\hat{\eta}(\omega)\int_0^t u(s)e^{\omega^2 s}\,ds + \hat{f}(\omega)\right\}, \qquad (2.40)$$

which can be inverted by (2.27) to give

$$T(x, t) = \frac{1}{\sqrt{2\pi}}\int_{-\infty}^{\infty} e^{-i\omega x}e^{-\omega^2 t}\hat{\eta}(\omega)\int_0^t u(s)e^{\omega^2 s}\,ds\,d\omega$$

$$+ \frac{1}{\sqrt{2\pi}}\int_{-\infty}^{\infty} e^{-i\omega x}e^{-\omega^2 t}\hat{f}(\omega)\,d\omega. \qquad (2.41)$$

Substituting this into (2.37) and exchanging order of integration, we obtain

$$\frac{du(t)}{dt} = \frac{1}{\sqrt{2\pi}}\int_0^t u(s)\int_{-\infty}^{\infty}\int_{-\infty}^{\infty} \alpha(x)e^{-i\omega x}e^{-\omega^2(t-s)}\hat{\eta}(\omega)\,dx\,d\omega\,ds$$

$$+ \frac{1}{\sqrt{2\pi}}\int_{-\infty}^{\infty}\int_{-\infty}^{\infty} \alpha(x)e^{-i\omega x}e^{-\omega^2 t}\hat{f}(\omega)\,dx\,d\omega. \qquad (2.42)$$

This can be written in the more explicit form

$$\frac{du(t)}{dt} = \int_0^t k(t,s)u(s)\,ds + g(t), \qquad (2.43)$$

where

$$k(t,s) = -\int_{-\infty}^{\infty} \hat{\alpha}(-\omega)\hat{\eta}(\omega)e^{-\omega^2(t-s)}\,d\omega,$$

and

$$g(t) = -\int_{-\infty}^{\infty} \hat{\alpha}(-\omega)\hat{f}(\omega)e^{-\omega^2 t}\,d\omega.$$

Equation (2.43) is a simple Volterra integrodifferential equation for a function of one variable.

2.4. Some problems in experimental inference. Since Volterra equations are generalizations of initial value problems for ordinary differential equations, it is not surprising that most applications occur in connection with time-dependent systems. This was the case in all examples in the previous sections of this chapter. There are, however, instances where Volterra equations occur in the description of completely static phenomena. One class of such applications involves the inference of quantities which are not directly observable, but which can be measured through certain "averaged" data.

Consider, for example, a three-dimensional radiating body B in which the emission coefficient f (that is, the amount of radiation emitted by an infinitesimal volume element) varies with x, y, and z. In such situations $f(x, y, z)$ is often unknown and has to be inferred from external measurements. Typically, the intensity of radiation is measured at various points outside B with the hope that, if enough measurements are taken, $f(x, y, z)$ can be computed. In general this leads to Fredholm equations of the first kind, but when radial symmetries are present these may reduce to equations of the Abel type.

Assume that B is a circular cylinder and consider the cross-section with the plane $z = 0$ (Fig. 2.3).

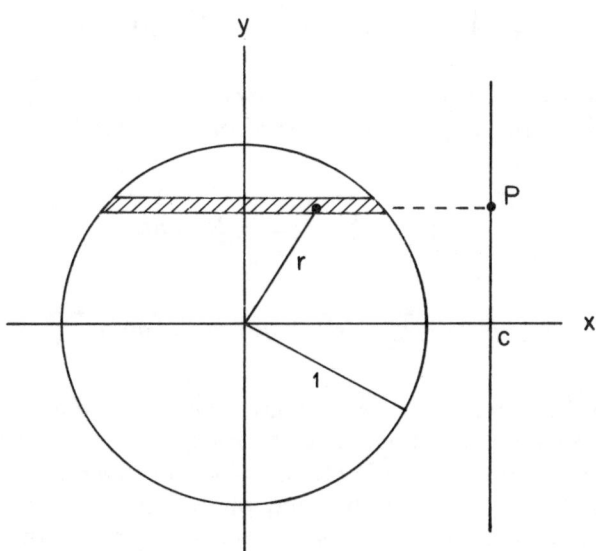

Fig. 2.3. *External measurement of radiation of a cylindrical object.*

At the point $P=(c, y)$ we measure the radiation in the x-direction, denoting it by $g(y)$. Then clearly

$$g(y) = \int_{-\sqrt{1-y^2}}^{\sqrt{1-y^2}} h(x, y) \, dx, \qquad (2.44)$$

where $h(x, y) = f(x, y, 0)$. If we further assume radial symmetry for $h(x, y)$, that is,

$$h(x, y) = R(r), \qquad (2.45)$$

where $r = \sqrt{x^2 + y^2}$, then

$$g(y) = \int_{-\sqrt{1-y^2}}^{\sqrt{1-y^2}} R(r) \, dx. \qquad (2.46)$$

A transformation of variables then gives the equation

$$g(y) = 2 \int_y^1 \frac{rR(r)}{\sqrt{r^2 - y^2}} \, dr, \qquad (2.47)$$

which is an equation of the Abel type for the unknown $R(r)$.

A slightly more complicated situation arises when B not only emits radiation, but also absorbs it. Radiation originating at any point inside B is attenuated when traveling through B; in general, this attenuation is exponential with the distance traveled. If, for simplicity, we take the coefficient of attenuation ρ as constant throughout B, then (2.44) must be replaced by

$$g(y) = \int_{-\sqrt{1-y^2}}^{\sqrt{1-y^2}} \exp\{-\rho(\sqrt{1-y^2} - x)\} h(x, y) \, dx, \qquad (2.48)$$

since radiation originating at (x, y) travels a distance $\sqrt{1-y^2} - x$ before leaving B. After the appropriate change of variables, (2.48) becomes the generalized Abel equation

$$g(y) = \int_y^1 \frac{rk(y, r)}{\sqrt{r^2 - y^2}} R(r) \, dr, \qquad (2.49)$$

where

$$k(y, r) = \exp\{-\rho(\sqrt{1-y^2} - \sqrt{r^2 - y^2})\} + \exp\{-\rho(\sqrt{1-y^2} + \sqrt{r^2 - y^2})\}. \qquad (2.50)$$

As a second example we take a situation which often arises in stereology and related fields. Spherical particles of various sizes are embedded in some material. The distribution of particle size is unknown and is to be inferred by taking slices through the medium and observing the particles within the slices. Generally, what can be measured is the apparent radius, which is the radius of that portion of a particle lying within the slice of thickness D (Fig. 2.4).

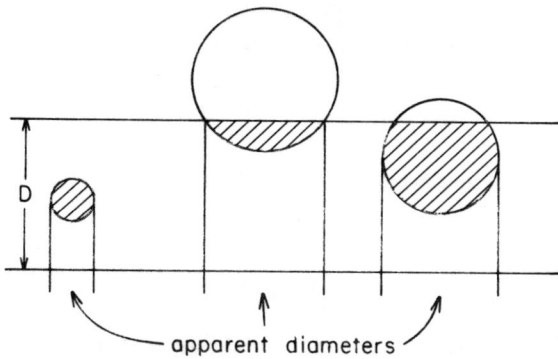

FIG. 2.4. *Apparent diameters of spherical particles embedded in a slice.*

For the sake of simplicity, assume that the radii are all in the range $0 \leq x \leq R$. Let ρ denote the density of particle centers and assume that it is constant throughout the material. The probability density for particle radii will be denoted by $f(x)$, so that

$$f(x)\,dx$$

is the probability that a particle has radius between x and $x + dx$. Lastly, assume that the slice has unit surface area.

We can then make the following simple arguments. If the center of a particle of radius x lies in the slice, then its apparent radius is x. If a particle of radius r has its center at a distance y from the surface of the slice, then its apparent radius is

$$x = \sqrt{r^2 - y^2}. \tag{2.51}$$

Thus, the number of particles with apparent radius between x and $x + dx$ is computed from two parts, the first being particles whose center is in the slice and whose radius is between x and $x + dx$. The total from all such particles in the slice is

$$\rho D f(x)\,dx. \tag{2.52}$$

The second contribution comes from particles whose centers are located at a distance y outside the slice. The apparent radius is between x and $x + dx$ only if its actual radius is between r and $r + dr$, where the relation between x and r is given by (2.51). In particular

$$dr = \frac{x\,dx}{r},$$

so that the contribution from particles at y is proportional to

$$f(r)\frac{x}{r}\,dx.$$

This is to be summed over all possible y, that is, $0 \leq y \leq \sqrt{R^2 - x^2}$ and $-\sqrt{R^2 - x^2} \leq y \leq 0$, giving a total contribution of

$$2\rho \int_0^{\sqrt{R^2-x^2}} f(r) \frac{x}{r} \, dx \, dy.$$

On changing the integration with respect to y to one with respect to r, this becomes

$$2\rho x \, dx \int_x^R \frac{f(r)}{\sqrt{r^2 - x^2}} \, dr. \tag{2.53}$$

If we now let $G(x) \, dx$ denote the observed number of particles with apparent radius between x and $x + dx$, then combining (2.52) and (2.53) we see that

$$Df(x) + 2x \int_x^R \frac{f(r)}{\sqrt{r^2 - x^2}} \, dr = \frac{1}{\rho} G(x), \quad 0 \leq x \leq R. \tag{2.54}$$

This is a singular Volterra equation of the second kind.

In some cases it is possible to find the apparent radii only if $D = 0$, that is, only the surface of the slice can be examined. In this case (2.54) becomes an equation of the Abel type

$$\int_x^R \frac{f(r)}{\sqrt{r^2 - x^2}} \, dr = \frac{1}{2x\rho} G(x). \tag{2.55}$$

The practical solution of the integral equations arising in this connection is complicated by the fact that the quantities observed are subject to experimental errors. The effect of these uncertainties on the inferred solution must be carefully considered. This is a point to which we will return in later chapters.

Notes on Chapter 2. There are many books and papers dealing with Volterra equations arising in practical applications. We give here only a few references which are representative of what exists and which give the reader an introduction to a vast literature.

Miller [190] and Noble [196] give elementary discussions of applications at much the same level as this chapter, showing the essence of the subject while omitting some details.

The original work on renewal equations was done by Lotka [173] and Feller [101]. Because of its importance, much work has been done on renewal equations. The books by Bellman [25] and Bellman and Cooke [26] contain a wealth of material. Some newer material is in [226] and [235].

The original work of Volterra can be found in [231]. Davis [75] is a more accessible reference. More recent work dealing with population growth can be found in [41] and [221].

A simple discussion of the relation between Volterra equations and system modeling is given in [105], while [71] presents a much more extensive and theoretical approach.

Mann and Wolf [179] discuss the heat conduction problem, while a discussion of the equations of nuclear reactor dynamics can be found in [159].

Edels, Hearne and Young [88], Minerbo and Levy [191], and Nestor and Olsen [193] discuss the Abel equation arising from the radiating sphere. There are many papers, for example [7], [8], [9], [119], [120], [143], [204], dealing with applications in stereology and related questions.

In addition to the applications mentioned in this chapter, Volterra equations occur in areas such as viscoelasticity [73], [86], [136], [216], study of epidemics [234], superfluidity [160], and damped vibrations [161].

Chapter 3

Linear Volterra Equations of the Second Kind

The simplest and most completely understood Volterra equation is the linear equation of the second kind,

$$f(t) = g(t) + \int_0^t k(t, s) f(s) \, ds. \tag{3.1}$$

In §§ 3.1 and 3.2 we look at the questions of existence and uniqueness of the solution and its formal representation by the use of resolvent kernels. Section 3.3 is devoted to the study of some qualitative properties of the solution including comparison theorems, its behavior as $t \to \infty$, and its sensitivity to perturbations. In §§ 3.4 to 3.6 some of the results are extended to equations with unbounded kernels, systems of equations, and integrodifferential equations.

3.1. Existence and uniqueness of the solution. The classical approach to proving the existence and uniqueness of the solution of (3.1) is the method of successive approximation, also called the *Picard* method. This consists of the simple iteration

$$f_n(t) = g(t) + \int_0^t k(t, s) f_{n-1}(s) \, ds, \quad n = 1, 2, \ldots, \tag{3.2}$$

with

$$f_0(t) = g(t).$$

For ease of manipulation it is convenient to introduce

$$\varphi_n(t) = f_n(t) - f_{n-1}(t), \quad n = 1, 2, \ldots, \tag{3.3}$$

with

$$\varphi_0(t) = g(t).$$

On subtracting from (3.2) the same equation with n replaced by $n-1$, we see that

$$\varphi_n(t) = \int_0^t k(t,s)\varphi_{n-1}(s)\,ds, \qquad n = 1, 2, \ldots. \tag{3.4}$$

Also, from (3.3),

$$f_n(t) = \sum_{i=0}^n \varphi_i(t). \tag{3.5}$$

The first theorem uses this iteration to prove the existence and uniqueness of the solution under quite restrictive conditions, namely that $k(t,s)$ and $g(t)$ are continuous.

THEOREM 3.1. *If $k(t,s)$ is continuous in $0 \leq s \leq t \leq T$ and $g(t)$ is continuous in $0 \leq t \leq T$, then the integral equation (3.1) possesses a unique continuous solution for $0 \leq t \leq T$.*

Proof. Choose G and K such that

$$|g(t)| \leq G, \qquad 0 \leq t \leq T,$$
$$|k(t,s)| \leq K, \qquad 0 \leq s \leq t \leq T.$$

We first prove by induction that

$$|\varphi_n(t)| \leq \frac{G(Kt)^n}{n!}, \qquad 0 \leq t \leq T, \quad n = 0, 1, \ldots. \tag{3.6}$$

If we assume that (3.6) is true for $n-1$, then from (3.4)

$$|\varphi_n(t)| \leq \frac{GK^n}{(n-1)!} \int_0^t s^{n-1}\,ds = \frac{GK^n t^n}{n!}.$$

Since (3.6) is obviously true for $n = 0$, it holds for all n. This bound makes it obvious that the sequence $f_n(t)$ in (3.5) converges and we can write

$$f(t) = \sum_{i=0}^\infty \varphi_i(t). \tag{3.7}$$

We now show that this $f(t)$ satisfies equation (3.1).

The series (3.7) is uniformly convergent since the terms $\varphi_i(t)$ are dominated by $G(KT)^i/i!$. Hence we can interchange the order of integration and summation in the following expression to obtain

$$\int_0^t k(t,s) \sum_{i=0}^\infty \varphi_i(s)\,ds = \sum_{i=0}^\infty \int_0^t k(t,s)\varphi_i(s)\,ds$$
$$= \sum_{i=0}^\infty \varphi_{i+1}(t) = \sum_{i=0}^\infty \varphi_i(t) - g(t).$$

This proves that $f(t)$ defined by (3.7) satisfies equation (3.1). Each of the $\varphi_i(t)$ is clearly continuous. Therefore $f(t)$ is continuous, since it is the limit of a uniformly convergent sequence of continuous functions.

To show that $f(t)$ is the only continuous solution, suppose there exists another continuous solution $\tilde{f}(t)$ of (3.1). Then

$$f(t) - \tilde{f}(t) = \int_0^t k(t, s)\{f(s) - \tilde{f}(s)\}\, ds. \tag{3.8}$$

Since $f(t)$ and $\tilde{f}(t)$ are both continuous, there exists a constant B such that

$$|f(t) - \tilde{f}(t)| \leq B, \quad 0 \leq t \leq T.$$

Substituting this into (3.8)

$$|f(t) - \tilde{f}(t)| \leq KBt, \quad 0 \leq t \leq T,$$

and repeating the step shows that

$$|f(t) - \tilde{f}(t)| \leq \frac{B(Kt)^n}{n!}, \quad 0 \leq t \leq T,$$

for any n. For large enough n, the right-hand side is arbitrarily small, so that we must have

$$f(t) = \tilde{f}(t), \quad 0 \leq t \leq T,$$

and there is only one continuous solution. $\square$

Note that although there is only one continuous solution, there may be solutions which are not continuous. One well-known example is the equation

$$f(t) = \int_0^t s^{t-s} f(s)\, ds, \tag{3.9}$$

which has one obvious solution $f(t) = 0$. The kernel s^{t-s} is continuous, but it can be verified that

$$f(t) = ct^{t-1} \tag{3.10}$$

is a solution of (3.9) for any c. For $c \neq 0$ this solution is discontinuous at $t = 0$. It is not difficult to show that any discontinuous solution of (3.1) must be nonintegrable and this is true of (3.10).

Exercise 3.1. Prove that any integrable solution of (3.1) must, under the conditions of Theorem 3.1, be continuous.

Exercise 3.2. Let $p(t)$ and $q(t)$ be continuous on $0 \leq t \leq T$ and such that $0 \leq p(t) \leq q(t) \leq t$. Prove that under the conditions of Theorem 3.1 the equation

$$f(t) = g(t) + \int_{p(t)}^{q(t)} k(t, s) f(s)\, ds, \qquad 0 \le t \le T,$$

has a unique continuous solution.

The Picard iteration, equation (3.2), is not restricted to continuous kernels and functions, but can also be carried out in general function spaces. The proof of Theorem 3.1 is a simple case of a *contraction mapping* argument. It is possible to repeat what has been done here in a more general setting and thereby obtain the result of Theorem 3.1 under less restrictive conditions. For example, the same kind of argument can be made for square-integrable kernels. However, we will not pursue this, since for our purposes here such general results are not needed. Instead, let us look at another way of proving existence and uniqueness. The conditions needed for the next theorem do arise in some cases of interest not covered by Theorem 3.1.

The approach used in the next result can be called the *method of continuation*. We first establish existence and uniqueness in some interval $[0, T_1]$, then show that this solution can be continued to successive intervals $[T_1, T_2]$, $[T_2, T_3]$, and so on. Under suitable conditions, we eventually cover the whole interval $[0, T]$.

THEOREM 3.2. *Assume that in* (3.1)
  (i) $g(t)$ *is continuous in* $0 \le t \le T$,
  (ii) *for every continuous function h and all* $0 \le \tau_1 \le \tau_2 \le t$ *the integrals*

$$\int_{\tau_1}^{\tau_2} k(t, s) h(s)\, ds$$

and

$$\int_0^t k(t, s) h(s)\, ds \tag{3.11}$$

are continuous functions of t,
  (iii) $k(t, s)$ *is absolutely integrable with respect to s for all* $0 \le t \le T$,
  (iv) *there exist points* $0 = T_0 < T_1 < T_2 < \cdots < T_N = T$ *such that, for all i and all* $T_i \le t \le T_{i+1}$

$$\int_{T_i}^{\min(t,\, T_{i+1})} |k(t, s)|\, ds \le \alpha < 1, \tag{3.12}$$

where α is independent of t and i,
  (v) *for every t in* $[0, T]$

$$\lim_{\delta \to 0^+} \int_t^{t+\delta} |k(t+\delta, s)|\, ds = 0. \tag{3.13}$$

Then (3.1) *has a unique continuous solution for* $0 \le t \le T$.

Proof. Consider first the interval $[0, T_1]$. Define $f_n(t)$ and $\varphi_n(t)$ as in (3.2) and (3.3), respectively. Then

$$\varphi_n(t) = \int_0^t k(t,s)\varphi_{n-1}(s)\,ds,$$

so that, by (3.12),

$$|\varphi_n(t)| \leq \alpha \max_{0 \leq \tau \leq T_1} |\varphi_{n-1}(\tau)|.$$

This can be applied several times to yield

$$\max_{0 \leq \tau \leq T_1} |\varphi_n(t)| \leq \alpha^n \max_{0 \leq \tau \leq T_1} |g(\tau)|.$$

Since $\alpha < 1$, we see that the sequence (3.5) is dominated by

$$\max_{0 \leq \tau \leq T_1} |g(\tau)| \sum_{i=0}^n \alpha^i < \max_{0 \leq \tau \leq T_1} |g(\tau)| \frac{1}{1-\alpha}$$

and converges uniformly. Also, because of (3.11), the $\varphi_n(t)$ are continuous functions and hence $f(t)$ defined by (3.7) is continuous.

To see that $f(t)$ is the only continuous solution, assume there exists another continuous solution $\tilde{f}(t)$. Then

$$f(t) - \tilde{f}(t) = \int_0^t k(t,s)\{f(s) - \tilde{f}(s)\}\,ds,$$

so that

$$|f(t) - \tilde{f}(t)| \leq \max_{0 \leq \tau \leq t} |f(\tau) - \tilde{f}(\tau)| \int_0^t |k(t,s)|\,ds, \qquad 0 \leq t \leq T_1,$$

which implies

$$\max_{0 \leq \tau \leq T_1} |f(\tau) - \tilde{f}(\tau)| \leq \alpha \max_{0 \leq \tau \leq T_1} |f(\tau) - \tilde{f}(\tau)|.$$

Since $\alpha < 1$ this can be true only if $f(t) = \tilde{f}(t)$, that is, the solution is unique.

Having established existence and uniqueness in $[0, T_1]$, we now proceed to the next interval $[T_1, T_2]$. We write the equation as

$$f(t) = G(t) + \int_{T_1}^t k(t,s)f(s)\,ds, \tag{3.14}$$

with

$$G(t) = g(t) + \int_0^{T_1} k(t,s)f(s)\,ds, \tag{3.15}$$

where, in (3.15), $f(s)$ is the solution obtained in the first step. But (3.14) is just a Volterra equation with origin shifted from 0 to T_1. We can therefore apply the basic step again. Note that, because of condition (ii), $G(t)$ is continuous, so that all assumptions of the theorem are satisfied for (3.14). Thus, (3.14) has a unique continuous solution in $[T_1, T_2]$, say $\psi(t)$. From (3.14) and condition (v) we see that

$$\lim_{t \to T_1^+} \psi(t) = G(T_1).$$

But obviously

$$\lim_{t \to T_1^-} f(t) = G(T_1),$$

so that the continuation from $f(t)$ to $\psi(t)$ is continuous and we have a unique continuous solution in $[0, T_2]$.

This argument can be repeated and since there is only a finite number of subintervals in $[0, T]$, we thereby construct the unique continuous solution in $[0, T]$. □

Again, this theorem can be generalized by repeating the above arguments in other function spaces.

Theorem 3.2 is more general than Theorem 3.1, since the conditions of Theorem 3.1 imply that assumptions (i) to (v) in Theorem 3.2 are automatically satisfied. On the other hand, the conditions of Theorem 3.2 allow some cases not covered by Theorem 3.1. For example, it is not necessary that $k(t, s)$ be continuous for (3.11) to hold, so that some types of discontinuous kernels can be analyzed by Theorem 3.2. Of greater interest are certain unbounded kernels, as shown by the following example.

Example 3.1. The kernel

$$k(t, s) = (t - s)^{-1/2}$$

satisfies the conditions of Theorem 3.2, so that the equation

$$f(t) = g(t) + \int_0^t (t-s)^{-1/2} f(s) \, ds$$

has a unique continuous solution for all continuous g.

We leave it as an exercise to show that this claim is true.

It is however not sufficient that $k(s, t)$ only satisfy assumptions (ii) and (iii) of Theorem 3.2. Something like the additional conditions is needed as can be seen from the next example.

Example 3.2. The equation

$$f(t) = \int_0^t (t^2 - s^2)^{-1/2} f(s) \, ds$$

has two continuous solutions, the trivial solution $f(t) = 0$ as well as the solution $f(t) = t$. The kernel $(t^2 - s^2)^{-1/2}$ satisfies conditions (ii) and (iii), but not the last two conditions of Theorem 3.2.

Exercise 3.3. Show that the kernels
$$k(t, s) = (t - s)^{-1/2}$$
and
$$k(t, s) = s(t^2 - s^2)^{-1/2}$$
satisfy the conditions (ii) to (v) of Theorem 3.2.

Exercise 3.4. Show that the kernel
$$k(t, s) = (t^2 - s^2)^{-1/2}$$
satisfies conditions (ii) and (iii), but not conditions (iv) and (v) of Theorem 3.2.

Exercise 3.5. Use Theorem 3.2 to show that the equation
$$f(t) = 1 + \int_0^t k(t-s)f(s)\, ds, \qquad 0 \leq t \leq T,$$
with
$$k(\tau) = \begin{cases} 1, & 0 \leq \tau \leq 1, \\ \frac{1}{2}, & \tau > 1, \end{cases}$$
has a unique continuous solution for all $0 \leq T < \infty$.

Exercise 3.6. Find the actual solution of the equation in Exercise 3.5.

3.2. The resolvent kernel. Let us take another look at the method of successive approximation, in particular, the computation of $\varphi_n(t)$ by (3.4). We have
$$\varphi_1(t) = \int_0^t k(t, s)g(s)\, ds,$$
$$\varphi_2(t) = \int_0^t k(t, s)\varphi_1(s)\, ds = \int_0^t k(t, s) \int_0^s k(s, \tau)g(\tau)\, d\tau\, ds.$$

If $k(t, s)$ and $g(t)$ are continuous, then the order of integration can be interchanged and we get
$$\varphi_2(t) = \int_0^t \int_\tau^t k(t, s)k(s, \tau)\, ds\, g(\tau)\, d\tau = \int_0^t k_2(t, \tau)g(\tau)\, d\tau,$$
where
$$k_2(t, \tau) = \int_\tau^t k(t, s)k(s, \tau)\, ds.$$

A similar result holds for the other $\varphi_n(t)$ and it follows immediately by induction that

$$\varphi_n(t) = \int_0^t k_n(t, s)g(s)\, ds, \qquad (3.16)$$

where

$$k_n(t, s) = \int_s^t k(t, \tau)k_{n-1}(\tau, s)\, d\tau, \qquad (3.17)$$

with $k_1(t, s) = k(t, s)$. The k_n are called the *iterated kernels*.

From (3.5) we then have

$$f_n(t) = g(t) + \int_0^t \Gamma_n(t, s)g(s)\, ds, \qquad (3.18)$$

where

$$\Gamma_n(t, s) = \sum_{i=1}^n k_i(t, s). \qquad (3.19)$$

If $k(t, s)$ is continuous and

$$|k(t, s)| \leq K, \qquad 0 \leq s \leq t \leq T,$$

then

$$|k_n(t, s)| \leq \frac{K^n(t-s)^{n-1}}{(n-1)!}. \qquad (3.20)$$

This follows immediately by induction on (3.17). Therefore

$$\Gamma(t, s) = \sum_{i=1}^\infty k_i(t, s) \qquad (3.21)$$

is uniformly convergent for $0 \leq s \leq t \leq T$. The function $\Gamma(t, s)$ is the *resolvent kernel* for $k(t, s)$.

THEOREM 3.3. *If $k(t, s)$ and $g(t)$ are continuous, then the unique continuous solution of (3.1) is given by*

$$f(t) = g(t) + \int_0^t \Gamma(t, s)g(s)\, ds. \qquad (3.22)$$

Proof. In

$$\int_0^t \Gamma(t, s)g(s)\, ds = \int_0^t \sum_{i=1}^\infty k_i(t, s)g(s)\, ds$$

the order of integration and summation can be interchanged. Therefore

$$\int_0^t \Gamma(t,s)g(s)\,ds = \sum_{i=1}^\infty \int_0^t k_i(t,s)g(s)\,ds$$

$$= \sum_{i=1}^\infty \varphi_i(t,s) = f(t) - g(t),$$

where the last step follows from (3.7). □

The resolvent kernel itself can be expressed as the solution of an integral equation.

THEOREM 3.4. *Under the assumptions of Theorem 3.3 the resolvent kernel $\Gamma(t,s)$ satisfies the equation*

$$\Gamma(t,s) = k(t,s) + \int_s^t k(t,\tau)\Gamma(\tau,s)\,d\tau, \qquad 0 \le s \le t \le T. \tag{3.23}$$

Proof. Using (3.21), we see that

$$\int_s^t k(t,\tau)\Gamma(\tau,s)\,d\tau = \int_s^t k(t,\tau)\sum_{i=1}^\infty k_i(\tau,s)\,d\tau$$

$$= \sum_{i=1}^\infty \int_s^t k(t,\tau)k_i(\tau,s)\,d\tau$$

$$= \sum_{i=1}^\infty k_{i+1}(t,s)$$

$$= \Gamma(t,s) - k(t,s). \qquad \square$$

Exercise 3.7. Sometimes it is convenient to write (3.17) and (3.23) in somewhat different forms. First show that the iterated kernels defined in (3.17) also satisfy

$$k_n(t,s) = \int_s^t k_p(t,\tau)k_{n-p}(\tau,s)\,d\tau,$$

for all $1 \le p \le n-1$. Use this to show that the resolvent kernel also satisfies the equation

$$\Gamma(t,s) = k(t,s) + \int_s^t \Gamma(t,\tau)k(\tau,s)\,d\tau.$$

Equation (3.23) involves an unknown function of two variables and is not generally useful for the actual computation of the resolvent kernel. In one important case, however, a simplification occurs that makes (3.23) computationally useful.

DEFINITION 3.1. If the kernel of (3.1) is a function of $t-s$ only, that is,

$$k(t,s) = k(t-s), \tag{3.24}$$

then k is said to be a *difference* kernel.

THEOREM 3.5. *If k is a difference kernel, and k(t) and g(t) are continuous, then the unique continuous solution of* (3.1) *is given by*

$$f(t) = g(t) + \int_0^t R(t-s)g(s)\,ds, \tag{3.25}$$

where the resolvent kernel $R(t)$ is the solution of

$$R(t) = k(t) + \int_0^t k(t-s)R(s)\,ds. \tag{3.26}$$

Proof. Let $R(t)$ be the unique continuous solution of (3.26). Consider then $f(t)$ defined by (3.25). Then

$$\int_0^t k(t-s)f(s)\,ds = \int_0^t k(t-s)\left\{g(s) + \int_0^s R(s-\tau)g(\tau)\,d\tau\right\}ds$$

$$= \int_0^t k(t-s)g(s)\,ds + \int_0^t \left\{\int_\tau^t k(t-s)R(s-\tau)\,ds\right\}g(\tau)\,d\tau,$$

where the last step involves the usual interchange of order of integration. If in the inner integral we substitute $s - \tau = u$, then, by (3.26),

$$\int_\tau^t k(t-s)R(s-\tau)\,ds = \int_0^{t-\tau} k(t-\tau-u)R(u)\,du$$

$$= R(t-\tau) - k(t-\tau).$$

Thus

$$\int_0^t k(t-s)f(s)\,ds = \int_0^t k(t-s)g(s)\,ds + \int_0^t \{R(t-\tau) - k(t-\tau)\}g(\tau)\,d\tau$$

$$= \int_0^t R(t-\tau)g(\tau)\,d\tau.$$

Applying now (3.25), we see that

$$\int_0^t k(t-s)f(s)\,ds = f(t) - g(t),$$

so that $f(t)$ defined by (3.25) is the solution of (3.1) when k is a difference kernel. □

For equations with difference kernels then, the resolvent kernel is itself the solution of an equation with a difference kernel. This observation will be useful when we take a closer look at such equations in Chapter 6.

3.3. Some qualitative properties of the solution. Even when an explicit solution of (3.1) cannot be found, it may be possible to obtain some qualitative information about it. Results on the smoothness of the solution, upper and lower bounds, and its asymptotic behavior are of practical importance.

In this section we summarize a few results which can be obtained by elementary means. These results are intended as a demonstration of the arguments rather than as a presentation of a coherent picture. In fact, for general kernels the situation seems quite complicated and no comprehensive theory exists. For equations with difference kernels, however, much more work has been done. Some of this will be outlined in Chapter 6; here we consider only the general case.

One concern of particular importance in the numerical solution is the smoothness, that is, the differentiability, of the solution. If we allow ourselves to impose sufficiently stringent conditions, the analysis is quite trivial. If in (3.1) the function g is continuously differentiable on $[0, T]$, $k(t, s)$ is bounded, and $(\partial/\partial t) k(t, s)$ is continuous, then (3.1) has a continuous solution and the equation can be differentiated to give

$$f'(t) = g'(t) + k(t, t)f(t) + \int_0^t \frac{\partial}{\partial t} k(t, s) f(s) \, ds.$$

Since $f(t)$ is continuous, it follows from this that $f'(t)$ is also continuous. If we assume further differentiability for $g(t)$ and $k(t, s)$, we can differentiate once more to get

$$f''(t) = g''(t) + k_0'(t)f(t) + k_0(t)f'(t)$$
$$+ k_1(t)f(t) + \int_0^t \frac{\partial^2}{\partial t^2} k(t, s) f(s) \, ds, \qquad (3.27)$$

where we use the notation

$$k_j(t) = \frac{\partial^j}{\partial \tau^j} k(\tau, s) \bigg|_{\tau = t, s = t}.$$

Since $f'(t)$ is continuous, we conclude that, if all derivatives needed in (3.27) are continuous, then $f''(t)$ is continuous. The extension of this argument is obvious.

THEOREM 3.6. *If*
  (i) $g(t)$ *is p times continuously differentiable in* $[0, T]$,
  (ii) $(\partial^j/\partial t^j) k(t, s)$ *is continuous in* $0 \leq s \leq t \leq T$ *for all* $j = 0, 1, \ldots, p$,
  (iii) $(d^q/dt^q) k_r(t)$ *is continuous in* $[0, T]$ *for all* $q \geq 0$ *and* $r \geq 0$ *such that* $r + q \leq p - 1$,
then the solution of (3.1) *is p times continuously differentiable in* $[0, T]$.

Proof. Inductively, carry on the process outlined above. We leave the details as exercise. □

Exercise 3.8. Complete the proof of Theorem 3.6.

Exercise 3.9. The rather stringent conditions of Theorem 3.6 can be relaxed in many ways. Show that Theorem 3.6 continues to hold if condition (ii) is replaced by:

(ii') *for every continuous function* $h(t)$ *and all* $j = 0, 1, \ldots, p$

$$\int_0^t \frac{\partial^j}{\partial t^j} k(t, s) h(s) \, ds$$

is a continuous function of t.

Apply this new result to investigate the smoothness of the solution of the equation

$$f(t) = 1 + \int_0^t \sqrt{t-s} f(s) \, ds.$$

The next set of theorems deals with bounds on the solution of (3.1) and its behavior as $t \to \infty$. The first is a simple comparison theorem useful for both theoretical and practical purposes.

THEOREM 3.7. *Assume that the kernel* $k(t, s)$ *in* (3.1) *is absolutely integrable with respect to s for all* $0 \le t \le T$ *and that the equation has a continuous solution. Assume also that there exist functions* $G(t)$ *and* $K(t, s)$ *satisfying*

$$|g(t)| < G(t), \qquad 0 \le t \le T, \tag{3.28}$$

$$|k(t, s)| < K(t, s), \qquad 0 \le s \le t \le T, \tag{3.29}$$

and such that the integral equation

$$F(t) = G(t) + \int_0^t K(t, s) F(s) \, ds \tag{3.30}$$

has a continuous solution $F(t)$ *for* $0 \le t \le T$. *Then*

$$|f(t)| < F(t), \qquad 0 \le t \le T. \tag{3.31}$$

Proof. From (3.1)

$$|f(t)| \le |g(t)| + \int_0^t |k(t, s)| |f(s)| \, ds$$

$$< G(t) + \int_0^t K(t, s) |f(s)| \, ds.$$

Subtracting this from (3.30) gives

$$F(t) - |f(t)| > \int_0^t K(t, s)\{F(s) - |f(s)|\} \, ds. \tag{3.32}$$

Since $F(0) - |f(0)| > 0$ and $K(t, s)$ is positive, it is clear that $F(t) - |f(t)| > 0$ for all $t \le T$, which is the required result. □

The use of this theorem requires that we find a $K(t, s)$ simple enough so that (3.30) can be solved. We illustrate the application of this result with a few examples.

Example 3.3. Let $g(t)$ and $k(t, s)$ be bounded by

$$|k(t, s)| < K, \qquad |g(t)| < G.$$

Then (3.30) becomes

$$F(t) = G + K \int_0^t F(s) \, ds,$$

which has the solution

$$F(t) = Ge^{Kt},$$

so that

$$|f(t)| < Ge^{Kt}, \qquad 0 \le t \le T.$$

Example 3.4. Let $g(t)$ and $k(t, s)$ be bounded as in Example 3.3. Let $\hat{f}(t)$ be an approximate solution to (3.1) such that

$$r(t) = g(t) + \int_0^t k(t, s) \hat{f}(s) \, ds - \hat{f}(t).$$

Then $f(t) - \hat{f}(t)$ satisfies

$$f(t) - \hat{f}(t) = r(t) + \int_0^t k(t, s) \{f(s) - \hat{f}(s)\} \, ds.$$

If $|r(t)| < \varepsilon$, then we see from the result of Example 3.3 that

$$|f(t) - \hat{f}(t)| < \varepsilon e^{Kt}.$$

The bounds established with the arguments in Examples 3.3 and 3.4 are often very crude. Sometimes much better results can be achieved if we use a $K(t, s)$ of the form

$$K(t, s) = P(t)Q(s)$$

and find the solution using the explicit form (1.27).

Example 3.5. Consider the equation

$$f(t) = \cos t + \int_0^t \cos^2 (t - s) e^{-(t-s)} f(s) \, ds.$$

Since the equation

$$F(t) = 1 + \int_0^t e^{-(t-s)} F(s) \, ds$$

can be solved by (1.27) to give

$$F(t) = 1 + t,$$

we see immediately that

$$|f(t)| \leq 1 + t.$$

The inequalities (3.28) and (3.29) cannot immediately be relaxed to

$$|g(t)| \leq G(t) \qquad (3.33)$$

and

$$|k(t, s)| \leq K(t, s). \qquad (3.34)$$

For Theorem 3.7 to hold under these weakened conditions, some further restrictions have to be imposed on the kernel. For example, if we assume that

$$\lim_{t \to 0} \int_0^t |k(t, s)| \, ds < 1, \qquad (3.35)$$

then the theorem holds with (3.28) and (3.29) replaced by (3.33) and (3.34) respectively.

Exercise 3.10. Prove that Theorem 3.7 holds if (3.35) is satisfied and (3.28) and (3.29) are replaced by (3.33) and (3.34), respectively.

Exercise 3.11. Use Example 3.2 to show that Theorem 3.7 does not hold if (3.28) and (3.29) are replaced by (3.33) and (3.34), but no additional requirements are made.

The next two theorems are concerned with kernels with certain sign constraints. They yield bounds as well as information on the asymptotic behavior of the solution. We will consider the special equation

$$f(t) = 1 + \int_0^t k(t, s) f(s) \, ds. \qquad (3.36)$$

Taking $g(t) = 1$ simplifies matters by eliminating difficulties due to varying g. Actually, (3.36) is not as special as may appear. At least for $g(t)$ that do not change sign in $[0, T]$, equation (3.1) can be transformed to (3.36). First, divide by $g(t)$, then introduce the new variable $u(t) = f(t)/g(t)$ and the kernel

$$\frac{k(t, s) g(s)}{g(t)}.$$

This gives the form (3.36). We will assume that the kernel in (3.36) is sufficiently smooth to guarantee a differentiable solution and permit all differentiations used in the proof of the following theorem.

THEOREM 3.8. *If in* (3.36), *we have*

$$k(t, s) \leq 0, \tag{3.37}$$

$$\frac{\partial}{\partial t} k(t, s) \geq 0, \tag{3.38}$$

for all $0 \leq s \leq t \leq T$, *then the solution of* (3.36) *satisfies*

$$0 \leq f(t) \leq 1. \tag{3.39}$$

Proof. From (3.36)

$$f'(t) = k(t, t)f(t) + \int_0^t \frac{\partial}{\partial t} k(t, s)f(s) \, ds. \tag{3.40}$$

Since we are assuming that k and its derivatives are bounded, we have that $f(0) = 1$ and $f'(0) \leq 0$. Assume now that t^* is the smallest positive number such that $f(t^*) = 0$. Then

$$f'(t^*) = \int_0^{t^*} \frac{\partial}{\partial t} k(t^*, s)f(s) \, ds. \tag{3.41}$$

Because of (3.38) and the fact that $f(t) \geq 0$ for $t \leq t^*$, the right-hand side of (3.41) is positive. However, at t^*, $f(t)$ changes from positive to negative, hence $f'(t^*)$ cannot be positive. This contradiction proves the left half of the inequality (3.39). The right part then follows easily from (3.36). □

THEOREM 3.9. *If the conditions of Theorem 3.8 hold, and if in addition*

$$\lim_{t \to \infty} \int_0^t k(t, s) \, ds = -\infty \tag{3.42}$$

and $\lim_{t \to \infty} f(t)$ *exists, then*

$$\lim_{t \to \infty} f(t) = 0. \tag{3.43}$$

Proof. Assume that

$$\lim_{t \to \infty} f(t) = a > 0.$$

We can then pick τ large enough so that $f(t) \geq a/2$ for all $t > \tau$. Then

$$\int_\tau^t k(t, s)f(s) \, ds \leq \frac{a}{2} \int_\tau^t k(t, s) \, ds$$

and

$$f(t) = 1 + \int_0^\tau k(t, s)f(s) \, ds + \int_\tau^t k(t, s)f(s) \, ds$$

$$\leq 1 + \int_0^\tau k(t, s)f(s) \, ds + \frac{a}{2} \int_\tau^t k(t, s) \, ds.$$

By (3.42), the last term on the right goes to $-\infty$ as $t \to \infty$, contradicting the fact that $f(t) \geq 0$. □

Exercise 3.12. Show that the solution of

$$f(t) = 1 - \int_0^t \frac{1}{\sqrt{1+t+s}} f(s)\, ds$$

satisfies

$$0 \leq f(t) \leq 1$$

and

$$\lim_{t \to \infty} f(t) = 0.$$

The question of the existence of $\lim_{t \to \infty} f(t)$ in Theorem 3.9 is left open. This matter could be settled easily if one could show that $f(t)$ decreased monotonically. Unfortunately, conditions (3.37) and (3.38) are not sufficient to guarantee this; nor do there seem to be any simple additional conditions, such as perhaps

$$\frac{\partial^2}{\partial t^2} k(t, s) \leq 0, \qquad (3.44)$$

that would do the trick. This can be seen from the following example.

Example 3.6. Consider

$$k(t, s) = \begin{cases} -(1-t)^3, & 0 \leq t \leq 1, \\ 0, & t > 1. \end{cases}$$

Then conditions (3.37) and (3.38), and (3.44) are all satisfied. But the solution of

$$f(t) = 1 - (1-t)^3 \int_0^t f(s)\, ds, \qquad 0 \leq t \leq 1,$$

is clearly less than 1 in $0 < t < 1$, while $f(t) = 1$ for $t \geq 1$.

Exercise 3.13. Show that if the conditions of Theorem 3.8 hold and if

$$\lim_{t \to \infty} \int_0^t k(t, s)\, ds = 0,$$

then $\lim_{t \to \infty} f(t)$ cannot be zero.

Even for the simplified equation (3.36) the characterization of the asymptotic behavior of the solution is difficult and contains a number of open problems. We will reconsider some of these questions in Chapter 6 for difference kernels.

As a final result, consider the behavior of the solution of (3.1) with respect to changes in g and k.

THEOREM 3.10. *Let $g(t)$, $k(t, s)$, $\Delta g(t)$, $\Delta k(t, s)$ be continuous and bounded by*

$$|k(t, s)| \leq K, \quad |\Delta k(t, s)| \leq \Delta K,$$
$$|g(t)| \leq G, \quad |\Delta g(t)| \leq \Delta G.$$

Let $\hat{f}(t)$ be the solution of

$$\hat{f}(t) = g(t) + \Delta g(t) + \int_0^t \{k(t, s) + \Delta k(t, s)\}\hat{f}(s) \, ds. \quad (3.45)$$

Then $\hat{f}(t)$ satisfies

$$|\hat{f}(t) - f(t)| \leq \{\Delta G + \Delta Kt(G + \Delta G)e^{(K + \Delta K)t}\}e^{Kt} \quad (3.46)$$
$$= O(\Delta G) + O(\Delta K), \quad (3.47)$$

where $f(t)$ is the solution of (3.1).

Proof. From the results of Example 3.3 we have immediately that

$$|\hat{f}(t)| \leq (G + \Delta G)e^{(K + \Delta K)t}.$$

Now substitute $\hat{f}(t)$ into (3.1). Then

$$r(t) = g(t) + \int_0^t k(t, s)\hat{f}(s) \, ds - \hat{f}(t)$$
$$= -\Delta g(t) - \int_0^t \Delta k(t, s)\hat{f}(s) \, ds,$$

so that

$$|r(t)| \leq \Delta G + \Delta Kt \max |\hat{f}(s)| \leq \Delta G + \Delta Kt(G + \Delta G)e^{(K + \Delta K)t}.$$

We now apply the result of Example 3.4 with

$$\varepsilon = \Delta G + \Delta Kt(G + \Delta G)te^{(K + \Delta K)t}$$

to yield (3.46). □

The order of magnitude estimate (3.47) for the effect of a perturbation shows that (3.1) is *well-posed*. By definition, a problem is said to be well-posed if it has a unique solution which depends continuously on the parameters, that is, small perturbations in the parameters cause a small corresponding change in the solution. Here the parameters are $k(t, s)$ and $g(t)$, and the size of the perturbations are measured by their norms, for example,

$$\|\Delta g\|_\infty = \max_{0 \leq t \leq T} |\Delta g(t)|,$$

with a similar definition for the norm of the perturbation in the kernel. A unique continuous solution of (3.1) is known to exist, and (3.46) shows its continuous dependence on $\|\Delta g\|$ and $\|\Delta k(t, s)\|$. Therefore, (3.1) is well-posed. The inequality (3.46) gives a little more information than (3.47) and indicates that if K is large, the sensitivity of the solution to a small change may be considerable. This is of course not necessarily so, since (3.46) is an inequality. But it does indicate that equations for which the kernel has large magnitude can be troublesome computationally.

3.4. Systems of equations. Some of the results just presented can be generalized to systems of equations with only minor changes in notation. Using the usual vector-matrix notation, the linear system

$$f_1(t) = g_1(t) + \int_0^t \sum_{i=1}^n k_{1i}(t, s) f_i(s)\, ds,$$

$$f_2(t) = g_2(t) + \int_0^t \sum_{i=1}^n k_{2i}(t, s) f_i(s)\, ds, \qquad (3.48)$$

$$\vdots$$

$$f_n(t) = g_n(t) + \int_0^t \sum_{i=1}^n k_{ni}(t, s) f_i(s)\, ds$$

can be written as

$$\mathbf{f}(t) = \mathbf{g}(t) + \int_0^t \mathbf{k}(t, s) \mathbf{f}(s)\, ds. \qquad (3.49)$$

Using the vector norm

$$\|\mathbf{f}(t)\| = \max_{1 \le i \le n} |f_i(t)|, \qquad (3.50)$$

the induced matrix norm is

$$\|\mathbf{k}(t, s)\| = \max_{1 \le i \le n} \sum_{j=1}^n |k_{ij}(t, s)|. \qquad (3.51)$$

With this notation, the steps involved in proving results for systems are often formally the same as those for single equations.

THEOREM 3.11. *If $\mathbf{g}(t)$ and $\mathbf{k}(t, s)$ are continuous in $0 \le s \le t \le T$ (meaning that all components are continuous), then the system (3.48) has a unique continuous solution for $0 \le t \le T$.*

Proof. Repeat the steps in the proof of Theorem 3.1, using norms in place of absolute values. □

In a similar vein, we can get analogues of the comparison result, Theorem 3.7.

THEOREM 3.12. *If the system (3.49) possesses a unique continuous solution*

$\mathbf{f}(t)$ in $0 \leq t \leq T$, such that $\mathbf{k}(t, s)\mathbf{f}(s)$ is absolutely integrable and if

$$\|\mathbf{g}(t)\| < G(t), \tag{3.52}$$

$$\|\mathbf{k}(t, s)\| < K(t, s) \tag{3.53}$$

with continuous K and G, then

$$\|\mathbf{f}(t)\| < F(t), \tag{3.54}$$

where $F(t)$ is the continuous solution of

$$F(t) = G(t) + \int_0^t K(t, s)F(s)\, ds. \tag{3.55}$$

Proof. Repeat the steps in the proof of Theorem 3.7, using norms instead of absolute values. □

Exercise 3.14. Carry out the details in the proofs of Theorems 3.11 and 3.12.

Exercise 3.15. Extend Theorem 3.10 to systems of equations.

3.5. Equations with unbounded kernels. In the previous sections of this chapter it was frequently assumed that $k(t, s)$ was bounded (often also continuous or smoother). To relax this condition is not a simple matter. Yet, equations with unbounded kernels are of practical interest; in particular, kernels containing factors $(t-s)^{-1/2}$ and $(t^2 - s^2)^{-1/2}$ arise in applications, as for example in (2.36).

While equations with unbounded kernels are on the whole not as well understood as those with smooth kernels, some of the previous results do carry over. A simple example is the equation

$$f(t) = g(t) + \int_0^t (t-s)^{-1/2} f(s)\, ds. \tag{3.56}$$

The method of successive substitutions can still be carried out and, if $g(t)$ is continuous, the result converges to a continuous function which is the solution of (3.56). To see this, compute the iterated kernel

$$k_2(t, s) = \int_s^t (t-\tau)^{-1/2}(\tau - s)^{-1/2}\, d\tau = \pi.$$

The second and higher iterated kernels are therefore bounded. Similar arguments can be made for $k(t, s) = (t-s)^{-\alpha}$, with $0 < \alpha < 1$. Eventually, some iterated kernel will be bounded, which then is used to show that the method of successive approximation converges. This argument requires, however, that we can manipulate the kernels to bound the iterated kernels. A more general approach is through the continuation arguments given in

Theorem 3.2. As we indicated there, the theorem can be applied to certain types of unbounded kernels.

When the kernel is unbounded (or has some other irregular behavior) it is often convenient to rewrite (3.1) as

$$f(t) = g(t) + \int_0^t p(t,s)k(t,s)f(s)\,ds, \tag{3.57}$$

where $p(t, s)$ represents the part with the nonsmooth behavior. This ill-behaved part is usually quite simple which helps in the analysis. Corresponding to Theorem 3.2 we then have the following result.

THEOREM 3.13. *Assume that in* (3.57)
  (i) $g(t)$ *is continuous in* $0 \leq t \leq T$,
  (ii) $k(t, s)$ *is continuous in* $0 \leq s \leq t \leq T$,
  (iii) *for each continuous function h and all* $0 \leq \tau_1 \leq \tau_2 \leq t$ *the integrals*

$$\int_{\tau_1}^{\tau_2} p(t,s)k(t,s)h(s)\,ds \tag{3.58}$$

and

$$\int_0^t p(t,s)k(t,s)h(s)\,ds$$

are continuous functions of t,
  (iv) $p(t, s)$ *is absolutely integrable with respect to s for all* $0 \leq t \leq T$,
  (v) *there exist points* $0 = T_0 < T_1 < T_2 < \cdots < T_N = T$ *such that with* $t \geq T_i$

$$K \int_{T_i}^{\min(t,\,T_{i+1})} |p(t,s)|\,ds \leq \alpha < 1, \tag{3.59}$$

where

$$K = \max_{0 \leq s \leq t \leq T} |k(t,s)|,$$

  (vi) *for every* $t \geq 0$

$$\lim_{\delta \to 0^+} \int_t^{t+\delta} |p(t+\delta, s)|\,ds = 0. \tag{3.60}$$

Then (3.57) *has a unique continuous solution in* $0 \leq t \leq T$.

Proof. Proceed as in Theorem 3.2. We sketch only the first step. Consider the interval $[0, T_1]$ and define $\varphi_n(t)$ as before. Then

$$\varphi_n(t) = \int_0^t p(t,s)k(t,s)\varphi_{n-1}(s)\,ds,$$

$$|\varphi_n(t)| \leq K \max_{0 \leq \tau \leq T_1} |\varphi_{n-1}(\tau)| \int_0^t |p(t,s)|\,ds,$$

and, by condition (v)

$$\max_{0\leq\tau\leq T_1} |\varphi_n(t)| \leq \alpha \max_{0\leq\tau\leq T_1} |\varphi_{n-1}(t)|,$$

with $\alpha < 1$. This leads to the conclusion that the equation has a unique continuous solution in $[0, T_1]$. This solution can then be shown to be continuable to the other intervals as in Theorem 3.2. □

Exercise 3.16. Complete the arguments in the proof of Theorem 3.13.

Exercise 3.17. Show that the equation

$$f(t) = 1 + \int_0^t \frac{e^{ts}}{\sqrt{t-s}} f(s)\, ds$$

has a unique continuous solution.

Exercise 3.18. Extend the results of Theorem 3.13 to systems of equations

$$\mathbf{f}(t) = \mathbf{g}(t) + \int_0^t p(t, s)\mathbf{k}(t, s)\mathbf{f}(s)\, ds,$$

where $p(t, s)$ is a scalar function satisfying the conditions given in Theorem 3.13.

3.6. Integrodifferential equations. The reduction of integrodifferential equations to a system of integral equations, described in § 1.1, can be used for the analysis of a variety of integrodifferential equations. Rather than develop a general theory, let us take a more specific case. The extension to equations of higher order is obvious.

THEOREM 3.14. *Consider the equation*

$$f'(t) = g(t) + h(t)f(t)$$
$$+ \int_0^t k_1(t, s)f(s)\, ds + \int_0^t k_2(t, s)f'(s)\, ds, \qquad (3.61)$$

with

$$f(0) = \alpha,$$

assuming that $g(t)$, $h(t)$, $k_1(t, s)$, $k_2(t, s)$ *are all continuous in* $0 \leq s \leq t \leq T$. *Then* (3.61) *has a unique continuously differentiable solution in* $0 \leq t \leq T$.

Proof. Let $f(t)$ and $z(t)$ be the solution of the system

$$z(t) = g(t) + h(t)f(t) \qquad (3.62)$$
$$+ \int_0^t k_1(t, s)f(s)\, ds + \int_0^t k_2(t, s)z(s)\, ds,$$
$$f(t) = \alpha + \int_0^t z(s)\, ds. \qquad (3.63)$$

Then, according to Theorem 3.11, the system (3.62), (3.63) has a continuous solution. Since $z(t)$ is continuous, we see from (3.63) that $f(t)$ is not only continuous, but also continuously differentiable with
$$f'(t) = z(t), \qquad f(0) = \alpha.$$
Therefore, $f(t)$ satisfies (3.61).

Since every solution of (3.61) satisfies a system of the form (3.62) and (3.63) and the latter has a unique continuous solution, the integrodifferential equation (3.61) must have a unique continuously differentiable solution. □

Other integrodifferential equations can be treated in a similar fashion.

Exercise 3.19. Show that the equation
$$f''(t) = g(t) + h_1(t)f'(t) + h_2(t)f(t)$$
$$+ \int_0^t k_1(t, s)f(s)\,ds + \int_0^t k_2(t, s)f'(s)\,ds + \int_0^t k_3(t, s)f''(s)\,ds,$$
$$f(0) = \alpha,$$
$$f'(0) = \beta,$$
with $g(t)$, $h_1(t)$, $h_2(t)$, $k_1(t, s)$, $k_2(t, s)$, $k_3(t, s)$ all continuous, has a unique twice continuously differentiable solution.

Exercise 3.20. Show that the equation
$$f'(t) = g(t) + \int_0^t \frac{f(s)}{\sqrt{t-s}}\,ds,$$
$$f(0) = \alpha,$$
with $g(t)$ continuous, has a unique continuously differentiable solution.

Notes on Chapter 3. The use of Picard iteration and resolvent kernels is standard and can be found in most texts on integral equations. The method of continuation, although more powerful than Picard iteration in demonstrating existence and uniqueness, seems to be less popular.

The setting used here, namely, the space of continuous functions, is commonly used. For more abstract approaches see Corduneanu [71] and Miller [190]. A treatment of equations with nonsmooth kernels can be found in Davis [74]; for unbounded kernels see [106].

As Theorem 3.6 shows, smoothness results for equations with differentiable kernels are quite trivial. This matter is considerably more complicated for important unbounded kernels such as $k(t, s) = (t-s)^{-\mu}$. For an investigation of this case, see [189].

Some elementary qualitative results, going beyond what is given here, can be found in [24], [125], and [208].

Chapter 4

Nonlinear Equations of the Second Kind

In this chapter we turn our attention to the nonlinear Volterra equation of the second kind

$$f(t) = g(t) + \int_0^t K(t, s, f(s))\, ds, \qquad 0 \leq t \leq T. \tag{4.1}$$

We investigate the solution of this equation under various assumptions on the kernel. In § 4.1 the existence and uniqueness results for linear equations are extended under the assumption that $K(t, s, u)$ satisfies a simple Lipschitz condition with respect to the third argument. While the analysis here is easy, the conditions are restrictive. In § 4.2 we consider some problems that arise in connection with kernels which are not Lipschitz continuous. Some qualitative properties of the solution are investigated in § 4.3. A brief discussion of equations with unbounded kernels and nonlinear systems is given in § 4.4. A discussion of an application of the resolvent kernel concept for nonlinear equations concludes the chapter.

4.1. Successive approximations for Lipschitz continuous kernels. The method of successive approximation as applied to nonlinear equations is a direct generalization of the method developed in Chapter 3 for the linear case. Successive iterates are defined by

$$f_n(t) = g(t) + \int_0^t K(t, s, f_{n-1}(s))\, ds, \tag{4.2}$$

with $f_0(t) = g(t)$. As in the linear case, we subtract from (4.2) a similar equation with n replaced by $n-1$. Then

$$f_n(t) - f_{n-1}(t) = \int_0^t \{K(t, s, f_{n-1}(s)) - K(t, s, f_{n-2}(s))\}\, ds. \tag{4.3}$$

To make the analysis as simple as possible we assume that $K(t, s, u)$ satisfies a *Lipschitz condition* of the form

$$|K(t, s, y) - K(t, s, z)| \leq L |y - z|, \tag{4.4}$$

where L is independent of t, s, y, and z.

Introducing as before

$$\varphi_n(t) = f_n(t) - f_{n-1}(t), \tag{4.5}$$

with $\varphi_0(t) = g(t)$, we see that

$$f_n(t) = \sum_{i=0}^{n} \varphi_i(t) \tag{4.6}$$

and

$$|\varphi_n(t)| \leq L \int_0^t |\varphi_{n-1}(s)| \, ds. \tag{4.7}$$

A theorem and proof for the nonlinear case, analogous to Theorem 3.1, is immediate.

THEOREM 4.1. *Assume that in* (4.1) *the functions* $g(t)$ *and* $K(t, s, u)$ *are continuous in* $0 \leq s \leq t \leq T$ *and* $-\infty < u < \infty$, *and that furthermore the kernel satisfies a Lipschitz condition of the form* (4.4). *Then* (4.1) *has a unique continuous solution for all finite T.*

Proof. From (4.7) it follows, as before, that

$$|\varphi_n(t)| \leq \frac{G(Lt)^n}{n!}, \tag{4.8}$$

where

$$G = \max_{0 \leq t \leq T} |g(t)|. \tag{4.9}$$

Therefore

$$f(t) = \sum_{i=0}^{\infty} \varphi_i(t) \tag{4.10}$$

exists and is a continuous function.

To prove that $f(t)$ defined by (4.10) satisfies the original equation, set

$$f(t) = f_n(t) + \Delta_n(t). \tag{4.11}$$

From (4.2)

$$f(t) - \Delta_n(t) = g(t) + \int_0^t K(t, s, f(s) - \Delta_{n-1}(s)) \, ds,$$

so that

$$f(t) - g(t) - \int_0^t K(t, s, f(s))\, ds$$
$$= \Delta_n(t) + \int_0^t \{K(t, s, f(s) - \Delta_{n-1}(s)) - K(t, s, f(s))\}\, ds. \quad (4.12)$$

Applying the Lipschitz condition (4.4) gives

$$\left| f(t) - g(t) - \int_0^t K(t, s, f(s))\, ds \right| \leq |\Delta_n(t)| + Lt\, \|\Delta_{n-1}\|, \quad (4.13)$$

where

$$\|\Delta_{n-1}\| = \max_{0 \leq s \leq t} |\Delta_{n-1}(s)|.$$

But

$$\lim_{n \to \infty} |\Delta_n(t)| = 0,$$

so that by taking n large enough, the right-hand side of (4.13) can be made as small as desired. It follows that the function $f(t)$ defined by (4.10) satisfies

$$f(t) = g(t) + \int_0^t K(t, s, f(s))\, ds,$$

and is therefore a solution of (4.1).

To show uniqueness, we assume the existence of another continuous solution $\tilde{f}(t)$. Then

$$f(t) - \tilde{f}(t) = \int_0^t \{K(t, s, f(s)) - K(t, s, \tilde{f}(s))\}\, ds, \quad (4.14)$$

from which it follows that

$$|f(t) - \tilde{f}(t)| \leq L \int_0^t |f(s) - \tilde{f}(s)|\, ds. \quad (4.15)$$

Since $|f(t) - \tilde{f}(t)|$ must be bounded, say by B, (4.15) implies that

$$|f(t) - \tilde{f}(t)| \leq BLt.$$

By repeating the argument we are led to

$$|f(t) - \tilde{f}(t)| \leq B \frac{(Lt)^n}{n!} \quad (4.16)$$

for any n, and hence conclude that $f(t) = \tilde{f}(t)$. $\square$

Exercise 4.1. Show that the equation
$$f(t) = 1 + \int_0^t \frac{\sin(t-s)}{1+f^2(s)}\,ds$$
has a unique continuous solution for all t.

4.2. Existence and uniqueness for more general kernels. The use of the Lipschitz condition (4.4) simplifies the analysis but makes the results of limited use. In many cases (4.4) is not satisfied, so that we need to examine more general nonlinear behavior.

To develop some intuition, let us consider some specific cases, for example, the equation
$$f(t) = 1 + \int_0^t f^2(s)\,ds. \tag{4.17}$$

In order to use arguments in Theorem 4.1, we would need a constant L such that
$$|y^2 - z^2| \leq L\,|y - z|,$$
that is,
$$|y + z| \leq L.$$
This is clearly impossible for arbitrary y and z, and (4.17) does not satisfy a condition of the form (4.4).

However, we note from the structure of the proof of Theorem 4.1 that the Lipschitz condition is applied only to bound the term
$$|K(t, s, f_n(s)) - K(t, s, f_{n-1}(s))|.$$
If we knew that $|f_n(t)| \leq B$ for all n, we could use $L = 2B$, and the proof would go through as before. Since
$$f(0) = f_n(0),$$
we expect that, at least in some neighborhood of the origin, the iterates $f_n(t)$ remain bounded. If we can show this, we will be able to prove existence and uniqueness in some region $0 \leq t \leq T$, with $T > 0$.

The actual solution of (4.17) can be found by differentiating and solving the resulting differential equation
$$f'(t) = f^2(t), \quad f(0) = 1, \tag{4.18}$$
which gives
$$f(t) = \frac{1}{1-t}. \tag{4.19}$$

A solution exists in $0 \le t < 1$, but not in any large interval; that is, existence and uniqueness holds only locally near $t = 0$.

In the general case, we can obtain similar results if we can show that there exists some interval $0 \le t \le \delta$, such that successive iterates remain bounded for $t \le \delta$.

THEOREM 4.2. *Consider* (4.1) *with* $g(t)$ *continuous in* $0 \le t \le T$. *Assume that there exist constants* α, β, L *such that*

 (i) $\alpha < g(t) < \beta$, $0 \le t \le T$,

 (ii) *for all* $0 \le s \le t \le T$ *and* $\alpha < u < \beta$, *the kernel* $K(t, s, u)$ *is continuous in all variables*,

 (iii) *for all* $0 \le s \le t \le T$ *and* $\alpha < y < \beta$, $\alpha < z < \beta$ *the kernel satisfies the Lipschitz condition*

$$|K(t, s, y) - K(t, s, z)| \le L |y - z|. \tag{4.20}$$

Then there exists a $\delta > 0$ *such that* (4.1) *has a unique continuous solution in* $0 \le t \le \delta$.

Proof. Because of (i) and (ii), we can find a constant C such that

$$\int_0^t |K(t, s, g(s))| \, ds \le Ct, \quad 0 \le t \le T. \tag{4.21}$$

Next, we choose a positive number d such that

$$\alpha < g(t) - Cde^{Ld} < g(t) < g(t) + Cde^{Ld} < \beta, \quad 0 \le t \le d. \tag{4.22}$$

This can always be done because of the strict inequalities in (i). Let

$$\delta = \min(d, T). \tag{4.23}$$

Defining f_n and φ_n as in (4.2) and (4.5), we can now show by induction that for $n = 1, 2, \ldots$

$$\alpha < f_n(t) < \beta, \tag{4.24}$$

$$|\varphi_n(t)| \le C \frac{L^{n-1} t^n}{n!}, \tag{4.25}$$

$$|f_n(t) - g(t)| \le C \sum_{i=1}^n \frac{L^{i-1} t^i}{i!}. \tag{4.26}$$

Assume that (4.24)–(4.26) hold for $1, 2, \ldots, n-1$. Then

$$|\varphi_n(t)| = |f_n(t) - f_{n-1}(t)|$$
$$= \left| \int_0^t K(t, s, f_{n-1}(s)) - K(t, s, f_{n-2}(s)) \, ds \right|$$
$$\le L \int_0^t |f_{n-1}(s) - f_{n-2}(s)| \, ds,$$

where the last step follows because f_{n-1} and f_{n-2} satisfy (4.24), so that the Lipschitz condition (4.20) can be applied. Thus

$$|\varphi_n(t)| \leq L \int_0^t |\varphi_{n-1}(s)| \, ds \leq C \frac{L^{n-1} t^n}{n!},$$

and (4.25) is satisfied. Also,

$$\begin{aligned}
|f_n(t) - g(t)| &= |f_{n-1}(t) - g(t) + \varphi_n(t)| \\
&\leq |f_{n-1}(t) - g(t)| + |\varphi_n(t)| \\
&\leq C \sum_{i=1}^{n-1} \frac{L^{i-1} t^i}{i!} + C \frac{L^{n-1} t^n}{n!}
\end{aligned}$$

and (4.26) holds for $|f_n(t) - g(t)|$. Finally, since for $0 \leq t \leq \delta$,

$$\sum_{i=1}^n \frac{L^{i-1} t^i}{i!} < d e^{Ld}, \tag{4.27}$$

(4.22) and (4.26) imply that

$$\alpha < f_n(t) < \beta. \tag{4.28}$$

For $n = 1$,

$$|f_1(t) - g(t)| = \left| \int_0^t K(t, s, g(s)) \, ds \right| \leq Ct,$$

so that (4.24) and (4.26) are satisfied. Also,

$$\varphi_1(t) = f_1(t) - g(t), \tag{4.29}$$

and (4.25) holds for $n = 1$. This completes the inductive argument proving (4.24)–(4.26).

The conclusion from this part of the argument is that in $0 \leq t \leq \delta$,

$$\alpha < f_n(t) < \beta,$$

for all n. This bounds the iterates and the rest of the proof is then completely analogous to that of Theorem 4.1. The details are left to the reader. □

Exercise 4.2. Complete the argument to prove Theorem 4.2.

Exercise 4.3. Show that for (4.17) all conditions of Theorem 4.2 are satisfied. Find a value of δ for which a continuous solution can be guaranteed to exist.

Exercise 4.4. Investigate the existence and uniqueness of the solution of

$$f(t) = 1 - \int_0^t \frac{e^{ts}}{f^2(s)} \, ds$$

near $t = 0$.

Theorem 4.2 is sufficiently general to guarantee the local existence and uniqueness of the solution under a wide variety of circumstances.

Example 4.1. Let

$$K(t, s, u) = k(t, s)h(u) \tag{4.30}$$

where $k(t, s)$ is continuous for $0 \leq s \leq t \leq T$ and $h(u)$ is continuously differentiable in some interval $\alpha = g(0) - c \leq u \leq g(0) + c = \beta$. Then Theorem 4.2 holds. To see this, note that

$$K(t, s, y) - K(t, s, z) = k(t, s)\{h(y) - h(z)\}, \tag{4.31}$$

so that

$$|K(t, s, y) - K(t, s, z)| \leq \max |k(t, s)| \, |y - z| \, |h'(\xi)|,$$

where $y \leq \xi \leq z$. We can therefore use

$$L = \max |k(t, s)| \max_{\alpha \leq \xi \leq \beta} |h'(\xi)| \tag{4.32}$$

to satisfy (4.20).

While Theorem 4.2 can be used to prove the existence and uniqueness of the solution near $t = 0$, it may yield fairly small values of δ, especially if rough estimates are used.

Example 4.2. Consider the solution of

$$f(t) = 1 - \int_0^t \frac{e^{ts}}{f(s)} \, ds \tag{4.33}$$

in $0 \leq t \leq 1$. If we pick $\alpha = \frac{1}{2}$, $\beta = 2$, then condition (i) and (ii) of Theorem 4.2 are satisfied. Also, for $0 \leq t \leq 1$,

$$\int_0^t e^{ts} \, ds \leq et,$$

so that (4.21) is satisfied with $C = e$.

With $\frac{1}{2} < y < 2$ and $\frac{1}{2} < z < 2$,

$$\left| \frac{e}{y} - \frac{e}{z} \right| \leq e \left| \frac{1}{y} - \frac{1}{z} \right| \leq e \left| \frac{z - y}{yz} \right| \leq 4e \, |y - z|,$$

so that L in (4.20) can be taken to be $4e$. Therefore, we must find a $d > 0$ such that

$$\tfrac{1}{2} < 1 - ede^{4ed} < 1 < 1 + ede^{4ed} < 2.$$

A value of $d = 0.07$ satisfies this and we conclude that (4.33) has a unique solution in $0 \leq t \leq 0.07$.

Exercise 4.5. Find a value for δ such that the equation

$$f(t) = t + \int_0^t \sin(t-s) f^3(s) \, ds$$

has a unique continuous solution in $0 \le t < \delta$.

4.3. Properties of the solution. The qualitative results developed for the linear case in § 3.3 can be extended to nonlinear equations if some additional assumptions are made. The smoothness result Theorem 3.6, is easily extended to the nonlinear case.

THEOREM 4.3. *If $g(t)$ is p times continuously differentiable in $[0, T]$ and $K(t, s, u)$ is p times continuously differentiable with respect to all three arguments in $0 \le s \le t \le T$, $-\infty < u < \infty$, then the solution of (4.1) is p times continuously differentiable.*

Proof. The proof immediately follows from successive differentiation of (4.1). □

To extend the comparison results and the theorems on the asymptotic behavior of the solution to nonlinear equations is rather difficult. The problem lies in the fact that not only do we need a characterization of the kernel (e.g., negative, increasing, etc.), but we must also say something about the type of nonlinearity. Some progress can be made if certain *monotonicity* properties are assumed. This is illustrated in the next set of theorems.

THEOREM 4.4. *Assume that the conditions of Theorem 4.1 hold. Let $f(t)$ be the continuous solution of (4.1) and let $F(t)$ be another continuous function satisfying*

$$F(t) = G(t) + \int_0^t H(t, s, F(s)) \, ds, \qquad (4.34)$$

where $G(t)$ and $H(t, s, u)$ are continuous in all arguments, and the following conditions hold:
  (i) $|g(t)| < G(t), 0 \le t \le T$,
  (ii) *for all functions $z_1(t)$, $z_2(t)$ such that*

$$|z_1(t)| \le z_2(t), \qquad (4.35)$$

the inequality

$$|K(t, s, z_1(t))| \le H(t, s, z_2(t)) \qquad (4.36)$$

holds for all $0 \le s \le t \le T$. Then

$$|f(t)| < F(t), \qquad 0 \le t \le T. \qquad (4.37)$$

Proof. Because of the continuity requirements, $|f(0)| < F(0)$. Now assume that (4.37) is not satisfied. Then there must exist some $\tau \le T$ such that

$|f(\tau)| = F(\tau)$. From (4.1) and (4.34)

$$F(t) - |f(t)| = G(t) + \int_0^t H(t, s, F(s)) \, ds$$
$$- \left| g(t) + \int_0^t K(t, s, f(s)) \, ds \right|$$
$$\geq G(t) - |g(t)| + \int_0^t \{H(t, s, F(s)) - |K(t, s, f(s))|\} \, ds.$$

Putting $t = \tau$ then gives

$$0 \geq G(\tau) - |g(\tau)| + \int_0^\tau \{H(\tau, s, F(s)) - |K(\tau, s, f(s))|\} \, ds. \qquad (4.38)$$

But $G(\tau) > |g(\tau)|$. Also, for $0 \leq t \leq \tau$, continuity implies that $F(t) \geq |f(t)|$. Therefore (4.36) shows that both parts of the right-hand side of (4.38) are positive. This contradiction proves (4.37). □

The assumption of strict inequality $|g(t)| < G(t)$ was made mainly to simplify the arguments. It can be removed without much difficulty.

Exercise 4.6. Show that if
 (i) $|g(t)| \leq G(t)$,
 (ii) $F(t)$ depends continuously on $G(t)$,
 (iii) the other conditions of Theorem 4.5 are satisfied,
then $|f(t)| \leq F(t)$ in $0 \leq t \leq T$.

Example 4.3. Consider the equation

$$f(t) = 1 + \int_0^t e^{-ts} \frac{1 + f(s) + f^3(s)}{1 + f^2(s)} \, ds. \qquad (4.39)$$

If we choose $G(t) = 1$ and

$$H(t, s, u) = 1 + |u|,$$

then the conditions of Theorem 4.4 (with Exercise 4.6) are satisfied. Since the equation

$$F(t) = 1 + \int_0^t \{1 + |F(s)|\} \, ds$$

obviously has a positive solution, we can omit the absolute value sign to get

$$F(t) = 1 + \int_0^t \{1 + F(s)\} \, ds.$$

This equation can be solved easily by differentiation to give

$$F(t) = 2e^t - 1.$$

Consequently, the solution $f(t)$ of (4.39) satisfies

$$f(t) \leq 2e^t - 1.$$

THEOREM 4.5. *Assume that the conditions of Theorem 4.1 are satisfied. Let $\hat{f}(t)$ be an approximate solution to equation (4.1) such that*

$$r(t) = g(t) + \int_0^t K(t, s, \hat{f}(s)) \, ds - \hat{f}(t). \tag{4.40}$$

Then

$$|f(t) - \hat{f}(t)| \leq Re^{Lt}, \tag{4.41}$$

where $f(t)$ is the solution of (4.1) and $R = \max |r(t)|$.

Proof. Subtracting (4.1) from (4.40) and using the Lipschitz condition (4.4), we get

$$|f(t) - \hat{f}(t)| \leq |r(t)| + L \int_0^t |f(s) - \hat{f}(s)| \, ds.$$

The result (4.41) then follows by comparing this equation with

$$F(t) = R + L \int_0^t F(s) \, ds. \qquad \square$$

The Lipschitz condition (4.4) is sufficient for the nonlinear problem to be well-posed.

Exercise 4.7. Let $f(t)$ be the solution of (4.1) and $\hat{f}(t)$ be the solution of

$$\hat{f}(t) = g(t) + \Delta g(t) + \int_0^t \{K(t, s, \hat{f}(s)) + \Delta K(t, s, \hat{f}(s))\} \, ds, \tag{4.42}$$

where Δg and ΔK are bounded for all $0 \leq s \leq t \leq T$, $-\infty < f < \infty$, and K and ΔK satisfy a Lipschitz condition of the form (4.4). Show that

$$|f(t) - \hat{f}(t)| = O(|\Delta g|) + O(|\Delta K|). \tag{4.43}$$

To get analogues of Theorems 3.8 and 3.9 we again take a simplified equation

$$f(t) = 1 + \int_0^t K(t, s, f(s)) \, ds. \tag{4.44}$$

As in the discussion preceding Theorem 3.8, the more general case can normally be reduced to (4.44).

THEOREM 4.6. *Let the conditions of Theorem 4.1 hold, and let $f(t)$ be the continuous solution of (4.44). Assume that the kernel satisfies the following* *conditions:*

(i) *for all u and* $0 \leq s \leq t \leq T$,

$$K(t, s, u) \geq 0 \quad \text{if } u \leq 0, \quad (4.45)$$

$$K(t, s, u) \leq 0 \quad \text{if } u \geq 0; \quad (4.46)$$

(ii) *for all* $u \geq 0$ *and* $0 \leq s \leq t \leq T$, *the function* $K(t, s, u)$ *is a nondecreasing function of t.*

Then

$$0 \leq f(t) \leq 1. \quad (4.47)$$

Proof. Since $f(0) = 1$ and $f(t)$ is a continuous function, there is some interval $0 \leq t \leq t_0$ for which $f(t) > 0$. But (4.46) then implies that $f(t) \leq 1$ for $0 \leq t \leq t_0$. If the solution were to change sign somewhere in $[0, T]$, there would be points t_1 and t_2 such that $f(t_1) = 0$ and $f(t) < 0$ for $t_1 < t \leq t_2$. Then, from (4.44)

$$0 > f(t_2) = 1 + \int_0^{t_1} K(t_2, s, f(s)) \, ds + \int_{t_1}^{t_2} K(t_2, s, f(s)) \, ds. \quad (4.48)$$

From assumption (ii)

$$K(t_2, s, f(s)) \geq K(t_1, s, f(s)), \quad 0 \leq s \leq t_1,$$

while from (4.45)

$$\int_{t_1}^{t_2} K(t_2, s, f(s)) \, ds \geq 0.$$

Therefore (4.48) becomes

$$0 > 1 + \int_0^{t_1} K(t_1, s, f(s)) \, ds = f(t_1) = 0.$$

This contradiction shows that $f(t)$ cannot change sign in $[0, T]$, thereby proving (4.47). □

THEOREM 4.7. *Assume that the conditions of Theorem 4.6 are satisfied. Furthermore, assume that for every* $a > 0$ *we have*

$$\lim_{t \to \infty} \int_0^t K(t, s, F(s)) \, ds = -\infty, \quad (4.49)$$

for every $F(t)$ *satisfying* $a \leq F(t) \leq 1$. *Then*

$$\lim_{t \to \infty} f(t) = 0, \quad (4.50)$$

if the limit exists.

Proof. The arguments are essentially the same as in Theorem 3.9. The crucial assumption is that (4.49) holds for every $F(t)$ bounded away from zero. □

Exercise 4.8. Complete the proof of Theorem 4.7.

Exercise 4.9. Show that the equation

$$f(t) = 1 - \int_0^t \frac{1}{\sqrt{1+t+s}} f^2(s)\, ds,$$

has a unique continuous solution for all $0 \le t < \infty$ satisfying

$$0 \le f(t) \le 1$$

and

$$\lim_{t \to \infty} f(t) = 0.$$

4.4. Unbounded kernels and systems of equations. The extension of the results in the foregoing sections is a relatively simple matter if we assume the stringent conditions of Theorem 4.1. For less restrictive conditions similar results can be obtained at the expense of more complicated arguments, but we will not pursue this matter here.

THEOREM 4.8. *Consider the equation*

$$f(t) = g(t) + \int_0^t p(t,s) K(t,s,f(s))\, ds, \tag{4.51}$$

where
  (i) *$g(t)$ is continuous in $0 \le t \le T$,*
  (ii) *$K(t, s, u)$ is a continuous function in $0 \le s \le t \le T$, $-\infty < u < \infty$,*
  (iii) *the Lipschitz condition*

$$|K(t, s, y) - K(t, s, z)| \le L |y - z| \tag{4.52}$$

is satisfied for $0 \le s \le t \le T$ and all y and z,
  (iv) *$p(t, s)$ satisfies conditions (iii)–(vi) of Theorem 3.13 with K replaced by L and $K(t, s, h(s))$ instead of $k(t, s)h(s)$.*

Then (4.51) has a unique continuous solution in $0 \le t \le T$.

Proof. Proceed as in the proof of Theorem 3.13. The Lipschitz condition (4.52) makes it possible to generate the same inequalities. ☐

Exercise 4.10. Complete the proof of Theorem 4.8.

THEOREM 4.9. *Consider the system of equations*

$$\mathbf{f}(t) = \mathbf{g}(t) + \int_0^t \mathbf{K}(t, s, \mathbf{f}(s))\, ds, \tag{4.53}$$

where we use the notation in § 4.1. Assume that
  (i) *$\mathbf{g}(t)$ is continuous (i.e., every component is continuous),*
  (ii) *$\mathbf{K}(t, s, \mathbf{u})$ is a continuous function for $0 \le s \le t \le T$ and $-\infty < \|\mathbf{u}\| < \infty$,*

(iii) *the kernel satisfies the Lipschitz condition*

$$\|\mathbf{K}(t, s, \mathbf{y}) - \mathbf{K}(t, s, \mathbf{z})\| \leq L \|\mathbf{y} - \mathbf{z}\|, \qquad (4.54)$$

where the norm is as defined in (3.50).
Then (4.53) *has a unique continuous solution in* $0 \leq t \leq T$.

Proof. Use the method of successive substitutions in Theorem 4.1, replacing absolute values with norms. All arguments hold as before. $\square$

Exercise 4.11. Complete the proof of Theorem 4.9.

One immediate application of Theorem 4.9 is in the study of integrodifferential equations.

Example 4.4. Consider the integrodifferential equation

$$f'(t) = \mathbf{H}\left(t, f(t), \int_0^t K(t, s, f(s)) \, ds\right), \qquad (4.55)$$

where H and K are continuous functions satisfying the Lipschitz conditions

$$|K(t, s, u_1) - K(t, s, u_2)| \leq L_1 |u_1 - u_2|, \qquad (4.56)$$

$$|H(t, v_1, w) - H(t, v_2, w)| \leq L_2 |v_1 - v_2|, \qquad (4.57)$$

$$|H(t, v, w_1) - H(t, v, w_2)| \leq L_3 |w_1 - w_2|. \qquad (4.58)$$

Integrating (4.55) and using $f(0) = \alpha$, we obtain the system

$$f(t) = \alpha + \int_0^t H(s, f(s), h(s)) \, ds, \qquad (4.59)$$

$$h(t) = \int_0^t k(t, s, f(s)) \, ds. \qquad (4.60)$$

Making the connection between H, K, f and h in (4.59) and (4.60) with the vectors $\mathbf{K}$, $\mathbf{y}$, $\mathbf{z}$ in (4.54), we see that

$$\begin{aligned}\|\mathbf{K}(t, s, \mathbf{y}) - \mathbf{K}(t, s, \mathbf{z})\| &= \max\{|H(t, y_1, y_2) - H(t, z_1, z_2)|, |K(t, s, y_1) - K(t, s, z_1)|\} \\ &\leq \max\{(L_2 |y_1 - z_1| + L_3 |y_2 - z_2|), L_1 |y_1 - z_1|\} \\ &\leq \max\{(L_2 + L_3, L_1)\} \max\{|y_1 - z_1|, |y_2 - z_2|\} \\ &\leq L \|\mathbf{y} - \mathbf{z}\|.\end{aligned}$$

Therefore, (4.54) is satisfied with

$$L = \max\{L_1, L_2 + L_3\}$$

and (4.55) has a unique continuous solution.

4.5. The resolvent equation. For nonlinear equations, there is no immediate analogue of Theorem 3.3, expressing the solution in terms of a resolvent kernel. Nevertheless, a resolvent equation can be used to establish a connection between certain nonlinear equations and related linear forms.

THEOREM 4.10. *Consider the nonlinear equation*

$$f(t) = g(t) + \int_0^t k(t,s)\{f(s) + H(s, f(s))\}\, ds. \tag{4.61}$$

Let $\Gamma(t, s)$ be the resolvent kernel for $k(t, s)$ as given in (3.21). Let $F(t)$ be defined by

$$F(t) = g(t) + \int_0^t \Gamma(t,s)g(s)\, ds. \tag{4.62}$$

Assume that all functions involved are continuous and such that (4.61) has a unique continuous solution on $[0, T]$ and $\Gamma(t, s)$ is a continuous function. Then $f(t)$ satisfies

$$f(t) = F(t) + \int_0^t \Gamma(t,s)H(s, f(s))\, ds. \tag{4.63}$$

Proof. Since we are assuming that (4.61) has a solution, we can write

$$f(t) = Q(t) + \int_0^t k(t,s)f(s)\, ds,$$

where

$$Q(t) = g(t) + \int_0^t k(t,s)H(s, f(s))\, ds.$$

Using now (3.22), we have

$$f(t) = Q(t) + \int_0^t \Gamma(t,s)Q(s)\, ds$$

$$= g(t) + \int_0^t k(t,s)H(s, f(s))\, ds$$

$$+ \int_0^t \Gamma(t,s)\left\{g(s) + \int_0^s k(s,\tau)H(\tau, f(\tau))\, d\tau\right\} ds$$

$$= g(t) + \int_0^t \Gamma(t,s)g(s)\, ds$$

$$+ \int_0^t H(\tau, f(\tau))\left\{k(t,\tau) + \int_\tau^t \Gamma(t,s)k(s,\tau)\, ds\right\} d\tau.$$

Applying the results of Exercise 3.7 gives

$$f(t) = g(t) + \int_0^t \Gamma(t,s)g(s)\, ds + \int_0^t \Gamma(t,\tau)H(\tau, f(\tau))\, d\tau.$$

Since this is the same as (4.62) and (4.63), the proof of the theorem is completed. □

Equations (4.62) and (4.63) are sometimes called the *variation of constants* formulas. When the nonlinearity in (4.61) is weak, that is, if the equation can be considered a small deviation from the linear case, variation of constants in conjunction with the method of successive approximations can yield some elementary perturbation results.

Example 4.5. Consider the equation

$$f(t) = g(t) + \int_0^t k(t,s)\{f(s) + \varepsilon H(s, f(s))\}\, ds, \qquad (4.64)$$

where ε is a small constant. Substituting this into (4.63) gives

$$f(t) = F(t) + \varepsilon \int_0^t \Gamma(t,s) H(s, f(s))\, ds.$$

Taking $f_0(t) = F(t)$, and substituting, we have

$$f_1(t) = F(t) + \varepsilon \int_0^t \Gamma(t,s) H(s, F(s))\, ds. \qquad (4.65)$$

If H is Lipschitz continuous in its second argument, then it is seen easily that

$$f(t) = f_1(t) + O(\varepsilon^2),$$

so that (4.65) gives the first-order perturbation effect of the nonlinearity.

Notes on Chapter 4. The use of the Lipschitz condition in Theorem 4.1, although restrictive, is simple and is used in most texts. Actually, it is not too difficult to relax the uniform condition (4.4) to

$$|K(t, s, y_1) - K(t, s, y_2)| \leq a(t, s)|y_1 - y_2|,$$

where $a(t, s)$ need only be, say, square integrable. For details see Tricomi [229]. A treatment of local existence and uniqueness for non-Lipschitz kernels is given in Corduneanu [70].

Sato [218] also discusses existence and uniqueness and gives some comparison theorems. A very extensive discussion and many results on nonlinear equations, including the variation of constants method, are in the book by Miller [190].

Chapter 5

Equations of the First Kind

Broadly speaking, there are two types of equations of the first kind which commonly occur in practical applications: (1) the kernel is a well-behaved (i.e., continuous, differentiable) function of all arguments, and (2) the kernel is unbounded at $s = t$. Abel's equation is an example in the second category.

In general, the theory of integral equations of the first kind is much less extensive than the theory for equations of the second kind. Consequently we will need to restrict our attention to special cases. In the first section we consider equations with smooth kernels with emphasis on the linear case. In the second section we consider Abel equations in various forms.

5.1. Equations with smooth kernels. We consider first the linear equation

$$\int_0^t k(t, s)f(s)\, ds = g(t), \qquad 0 \le t \le T. \tag{5.1}$$

If the kernel and the solution are to be bounded, then clearly it is necessary that $g(0) = 0$; otherwise no solution to (5.1) can exist.

Formally, (5.1) can be converted into an equation of the second kind by differentiation. If we differentiate (5.1) with respect to t, we obtain

$$k(t, t)f(t) + \int_0^t \frac{\partial k(t, s)}{\partial t} f(s)\, ds = g'(t). \tag{5.2}$$

If $k(t, t)$ does not vanish in $0 \le t \le T$, then, dividing (5.2) by $k(t, t)$ yields a standard Volterra equation of the second kind. The question of the existence and uniqueness of the solution of (5.1) can then be answered by considering (5.2).

THEOREM 5.1. *Assume that*
  (i) $k(t, s)$ *and* $\partial k(t, s)/\partial t$ *are continuous in* $0 \le s \le t \le T$,
  (ii) $k(t, t)$ *does not vanish anywhere in* $0 \le t \le T$,
  (iii) $g(0) = 0$,
  (iv) $g(t)$ *and* $g'(t)$ *are continuous in* $0 \le t \le T$.

Then (5.1) *has a unique continuous solution. This solution is identical with the continuous solution of* (5.2).

Proof. Since $k(t, t)$ does not vanish anywhere in $0 \leq t \leq T$, equation (5.2) is equivalent to the equation

$$f(t) + \int_0^t H(t, s) f(s) \, ds = G(t), \tag{5.3}$$

where

$$H(t, s) = \frac{1}{k(t, t)} \frac{\partial k(t, s)}{\partial t}, \qquad G(t) = \frac{g'(t)}{k(t, t)}.$$

Thus, for (5.3), all conditions of Theorem 3.1 are satisfied and there exists a unique continuous $f(t)$ satisfying (5.3) and (5.2). But (5.2) can be written as

$$\frac{d}{dt}\left\{\int_0^t k(t, s) f(s) \, ds - g(t)\right\} = 0.$$

Integrating this, and using the assumption $g(0) = 0$, shows that $f(t)$ satisfies (5.1). Also, since any solution of (5.1) must satisfy (5.2), equation (5.1) can have only one continuous solution. □

An alternative method for reducing (5.1) to an integral equation is to integrate by parts. With

$$\varphi(t) = \int_0^t f(s) \, ds, \tag{5.4}$$

we obtain from (5.1)

$$k(t, t)\varphi(t) - \int_0^t \frac{\partial k(t, s)}{\partial s} \varphi(s) \, ds = g(t). \tag{5.5}$$

This form indicates that Theorem 5.1 can be rewritten using somewhat different conditions on $k(t, s)$ and $g(t)$. We will leave this as an exercise.

Exercise 5.1. (a) Show that if condition (i) in Theorem 5.1 is replaced by
  (i) $k(t, s)$ and $\partial(t, s)/\partial s$ are continuous in $0 \leq s \leq t \leq T$,
then the conclusion of Theorem 5.1 still holds.

(b) Show that if $g(t)$ is continuous and $g'(t)$ bounded with possible simple jump discontinuities, then equation (5.1) has a unique bounded, but not necessarily continuous, solution.

The condition that $k(t, t)$ may not vanish at any point in $[0, T]$ is essential. Consider, for example,

$$\int_0^t (2t - 3s) f(s) \, ds = 0, \tag{5.6}$$

which violates the condition at the single point $t = 0$. By direct substitution it is easily verified that

$$f(t) = ct$$

satisfies (5.6) for any constant c, so that the equation has an infinite number of continuous solutions.

Exercise 5.2. Show that if $k(t, t) = 0$ for all $0 \leq t \leq T$, and if $k(t, s)$ and $g(t)$ are sufficiently differentiable, then (5.1) is equivalent to an equation of the second kind. What conditions must $g(t)$ satisfy for (5.1) to have a unique continuous solution under these circumstances?

Most of our subsequent discussion will be limited to linear equations. However, a brief comment on the nonlinear case

$$\int_0^t K(t, s, f(s)) \, ds = g(t) \tag{5.7}$$

is in order. The situation here is complicated by the fact that no solution may exist even for some rather simple examples. The equation

$$\int_0^t |f(s)| \, ds = -t \tag{5.8}$$

obviously has no solution, even though the kernel is constant with respect to t and s, and Lipschitz-continuous with respect to the third argument. The difficulty also appears when we try to convert (5.7) into an equation of the second kind. Differentiation of (5.7) gives

$$K(t, t, f(t)) + \int_0^t \frac{\partial K(t, s, f(s))}{\partial t} \, ds = g'(t), \tag{5.9}$$

which is not a standard equation of the second kind.

A set of conditions sufficient to guarantee a unique continuous solution is given in the following theorem.

THEOREM 5.2. *Assume that*

(i) $K(t, s, u)$ *and* $\partial K(t, s, u)/\partial t$ *are continuous for all* $0 \leq s \leq t \leq T, -\infty < u < \infty$,

(ii) *the function* $\partial K(t, s, u)/\partial t$ *satisfies a Lipschitz condition of the form*

$$\left| \frac{\partial K(t, s, y)}{\partial t} - \frac{\partial K(t, s, z)}{\partial t} \right| \leq L |y - z|, \tag{5.10}$$

for all y *and* z, *and* $0 \leq s \leq t \leq T$,

(iii) *the nonlinear equation*

$$K(t, t, x) = y \tag{5.11}$$

has a unique solution x *for every* y *and all* $0 \leq t \leq T$,

(iv) *there exists a constant* $\theta > 0$ *such that*
$$|K(t, t, y) - K(t, t, z)| \geq \theta |y - z|$$
for all y and z, and $0 \leq t \leq T$,
 (v) $g(0) = 0$,
 (vi) $g(t)$ *and* $g'(t)$ *are continuous in* $0 \leq t \leq T$.
Then (5.7) *has a unique continuous solution.*

Proof. We proceed as in Theorem 5.1. We consider the differentiated form (5.9), showing that it has a unique continuous solution, then conclude that this solution satisfies (5.7) and that it is the only solution of (5.7). The only thing new is to show that the nonstandard equation of the second kind (5.9) has a unique solution.

Consider the method of successive approximations applied to (5.9)
$$K(t, t, f_n(t)) = g'(t) - \int_0^t \frac{\partial K(t, s, f_{n-1}(s))}{\partial t} \, ds, \qquad n = 1, 2, \ldots, \qquad (5.12)$$
with some arbitrarily chosen continuous function $f_0(s)$. We want to show that the sequence $f_n(t)$ converges to a continuous function which is the solution of (5.9). To do so we can use the method of continuation as in Theorem 3.2. We will only sketch the required steps.

Because of condition (iii), for every given continuous $f_{n-1}(t)$, equation (5.12) uniquely defines $f_n(t)$. Because of condition (iv) and the various assumptions on K and g, $f_n(t)$ is continuous. Thus, (5.12) defines a unique sequence of continuous functions. Applying now (5.10) to (5.12), we get
$$|K(t, t, f_n(t)) - K(t, t, f_{n-1}(t))| \leq Lt \max_{0 \leq s \leq t} |f_{n-1}(s) - f_{n-2}(s)|. \qquad (5.13)$$

Using condition (iv) again, we have
$$|f_n(t) - f_{n-1}(t)| \leq \frac{Lt}{\theta} \max_{0 \leq s \leq t} |f_{n-1}(s) - f_{n-2}(s)|.$$

If we restrict t to the interval $[0, T_1]$ such that $LT_1/\theta < 1$, the existence and uniqueness of the solution follows by the usual contraction mapping arguments. From here we proceed as in Theorem 3.2, showing that there exists a continuous solution in $[T_1, 2T_1]$, and so on. □

Conditions (i), (ii) and (vi) of Theorem 5.2 can be relaxed in various ways. Condition (iii) is needed to eliminate such obvious counterexamples as (5.8). Condition (iv) plays the same role as does $k(t, t) \neq 0$ in the linear case.

Exercise 5.3. Complete all steps in the proof of Theorem 5.2.

Exercise 5.4. Extend Theorem 5.2 to the case where $\partial K(t, s, u)/\partial t$ and $g'(t)$ satisfy conditions similar to those in Theorem 4.2.

Exercise 5.5. Show that the equation

$$\int_0^t (1+t-s)^2\{f(s)+f^3(s)\}\,ds = t^2$$

has a unique continuous solution for all $t \geq 0$.

Qualitative properties of the solution of (5.1) can be studied using either of the equivalent forms (5.2) or (5.5) and various theorems, analogous to those in § 3.3, can be established. Not everything goes through trivially. In particular, the perturbation result given in Theorem 3.10, does not hold for equations of the first kind. Why this is so is easily seen from (5.2). The solution $f(t)$ depends, amongst other things, on the derivative of $g(t)$. If we measure the size of a function by the maximum norm, that is, for any continuous $\varphi(t)$, we define

$$\|\varphi\| = \max_{0 \leq t \leq T} |\varphi(t)|,$$

then we can make a small change in $g(t)$ which involves a large change in $g'(t)$, for example, by adding to $g(t)$ the function

$$\varepsilon \sin(t/\varepsilon)$$

whose magnitude is of order ε, but whose derivative is of order unity. Similarly, small changes in the kernel can have a significant effect on the solution as well. One way of expressing this observation is to say that (5.1) is not well-posed. It is a common observation that problems which are not well-posed can be troublesome numerically. Volterra equations of the first kind are no exception. We will return to this point in Chapter 9.

5.2. Abel equations. The general case of equations of the first kind with unbounded kernels is difficult and very little work has been done on it. Fortunately, practical situations often lead to kernels in which the singularity is of the form $(t-s)^{-\mu}$, $0 < \mu < 1$.

We begin with the simple Abel equation

$$\int_0^t \frac{f(s)}{(t-s)^\mu}\,ds = g(t), \qquad 0 \leq t \leq T. \tag{5.14}$$

As we will see, this equation has an explicit solution. Let us first take the case $\mu = \frac{1}{2}$, and consider

$$\int_0^t \frac{f(s)}{\sqrt{t-s}}\,ds = g(t). \tag{5.15}$$

In order to solve this equation, multiply both sides by $(x-t)^{-1/2}$ and integrate from 0 to x. Then

$$\int_0^x \frac{1}{\sqrt{x-t}} \int_0^t \frac{f(s)}{\sqrt{t-s}} \, ds \, dt = \int_0^x \frac{g(t)}{\sqrt{x-t}} \, dt. \tag{5.16}$$

Interchanging orders of integration on the left, we have

$$\int_0^x f(s) \int_s^x \frac{dt}{\sqrt{x-t}\sqrt{t-s}} \, ds = \int_0^x \frac{g(t)}{\sqrt{x-t}} \, dt. \tag{5.17}$$

The inner integral on the left can be evaluated explicitly

$$\int_s^x \frac{dt}{\sqrt{x-t}\sqrt{t-s}} = \int_0^{x-s} \frac{du}{\sqrt{u(x-s-u)}} = 2\int_0^1 \frac{dv}{\sqrt{1-v^2}} = \pi. \tag{5.18}$$

Consequently, from (5.17)

$$\pi \int_0^x f(s) \, ds = \int_0^x \frac{g(t)}{\sqrt{x-t}} \, dt, \tag{5.19}$$

or, on differentiating,

$$f(x) = \frac{1}{\pi} \frac{d}{dx} \int_0^x \frac{g(t)}{\sqrt{x-t}} \, dt. \tag{5.20}$$

To justify these manipulations and at the same time solve (5.14) for general μ, we need two standard results. First, for $0 \le \mu < 1$,

$$\int_t^x \frac{ds}{(x-s)^{1-\mu}(s-t)^\mu} = \int_0^{x-t} \frac{du}{(x-t-u)^{1-\mu} u^\mu}$$
$$= \int_0^1 (1-w)^{\mu-1} w^{-\mu} \, dw = \Gamma(\mu)\Gamma(1-\mu) = \frac{\pi}{\sin \mu\pi}. \tag{5.21}$$

Here Γ denotes the gamma function. The second result relates to the interchange of integrals. In general, it can be shown that if $\varphi(t, s)$ is a continuous function in $0 \le s \le t \le x$ and α, β, γ are constants such that $0 \le \alpha < 1$, $0 \le \beta < 1$, $0 \le \gamma < 1$ then

$$\int_0^x \int_0^t \frac{\varphi(t,s)}{(t-s)^\alpha s^\gamma (x-t)^\beta} \, ds \, dt = \int_0^x \int_s^x \frac{\varphi(t,s)}{(t-s)^\alpha s^\gamma (x-t)^\beta} \, dt \, ds. \tag{5.22}$$

Exercise 5.6. Prove that the interchange of order of integration in (5.22) is justified.

We can now find a closed form solution of (5.14).

THEOREM 5.3. *If $g(t)$ is continuous in $0 < t \leq T$ and*

$$\lim_{t \to 0} t^\alpha g(t) = C, \tag{5.23}$$

where $C \neq 0$ and $\alpha < \mu$, then (5.14) has the solution

$$f(t) = \frac{\sin \mu \pi}{\pi} \frac{d}{dt} \int_0^t \frac{g(s)}{(t-s)^{1-\mu}} \, ds, \qquad 0 < t \leq T. \tag{5.24}$$

This solution is continuous in $0 < t \leq T$, and satisfies

$$f(t) = \{C + o(1)\} \frac{\Gamma(1-\alpha)}{\Gamma(1-\mu)\Gamma(\mu-\alpha)} t^{\mu-\alpha-1}, \tag{5.25}$$

as $t \to 0$. Furthermore, this solution is unique in the class of functions of the form $f(t) = t^\beta F(t)$, where $\beta > -1$ and $F(t)$ is continuous.

Proof. Suppose first that (5.14) has a solution in the class of functions just stated. Multiply (5.14) by $(x-t)^{\mu-1}$ and integrate from 0 to x. Then

$$\int_0^x \int_0^t \frac{f(s)}{(x-t)^{1-\mu}(t-s)^\mu} \, ds \, dt = \int_0^x \frac{g(t)}{(x-t)^{1-\mu}} \, dt.$$

Since $f(s)$ is assumed to be of the form $s^\beta F(s)$, with $\beta > -1$ and $F(s)$ continuous, we can apply (5.22) to interchange order of integration and get

$$\int_0^x \int_s^x \frac{dt}{(x-t)^{1-\mu}(t-s)^\mu} f(s) \, ds = \int_0^x \frac{g(t)}{(x-t)^{1-\mu}} \, dt.$$

By (5.21) then

$$\frac{\pi}{\sin \mu \pi} \int_0^x f(s) \, ds = \int_0^x \frac{g(t)}{(x-t)^{1-\mu}} \, dt. \tag{5.26}$$

From (5.23)

$$\int_0^x \frac{g(t)}{(x-t)^{1-\mu}} \, dt \to C \int_0^x \frac{dt}{(x-t)^{1-\mu} t^\alpha} = C \frac{\Gamma(\mu)\Gamma(1-\alpha)}{\Gamma(1+\mu-\alpha)} x^{\mu-\alpha}$$

as $x \to 0$. Hence (5.26) can be differentiated everywhere except at $x = 0$. The results (5.24) and (5.25) then follow immediately.

We still must show that a solution in the stated class exists. This can be done by verification. Substituting (5.24) into the left-hand side of (5.14), we find, after a few manipulations, that it does indeed satisfy the equation. □

Exercise 5.7. Show that $f(t)$ given by (5.24) satisfies (5.14). Hint: Show first that if $G(t)$ is continuous and $\gamma > 0$, then

$$\frac{d}{dt}\int_0^t (t-s)^{-\mu} s^\gamma G(s)\, ds = \int_0^t (t-s)^{-\mu} \frac{d}{ds}\{s^\gamma G(s)\}\, ds.$$

Abel equations appear in the literature in various forms which are simply related to (5.14). For example, making the transformation of variables $t = u^2$, $s = v^2$, $F(v) = f(v^2)$, $G(u) = \frac{1}{2}g(u^2)$, equation (5.14) becomes

$$\int_0^u \frac{vF(v)}{(u^2-v^2)^\mu}\, dv = G(u). \qquad (5.27)$$

If, in (5.14), we set $t = 1-u$, $s = 1-v$, $F(v) = f(1-v)$, $G(u) = g(1-u)$, the equation reduces to

$$\int_u^1 \frac{F(v)}{(v-u)^\mu}\, dv = G(u). \qquad (5.28)$$

From this, by putting $u = z^2$, $v = w^2$, $\varphi(w) = F(w^2)$, $\xi(z) = \frac{1}{2}G(z^2)$, we get the form

$$\int_z^1 \frac{w\varphi(w)}{(w^2-z^2)^\mu}\, dw = \xi(z). \qquad (5.29)$$

Referring back to (5.24) the solutions to (5.26), (5.27), and (5.28) are easily found.

Exercise 5.8. (a) Show that the solution of (5.27) is given by

$$uF(u) = \frac{2\sin\mu\pi}{\pi}\frac{d}{du}\int_0^u \frac{vG(v)}{(u^2-v^2)^{1-\mu}}\, dv. \qquad (5.30)$$

(b) Show that the solution of (5.28) is given by

$$F(u) = -\frac{\sin\mu\pi}{\pi}\frac{d}{du}\int_u^1 \frac{G(v)}{(v-u)^{1-\mu}}\, dv. \qquad (5.31)$$

(c) Show that the solution of (5.29) is given by

$$z\varphi(z) = -\frac{2\sin\mu\pi}{\pi}\frac{d}{dz}\int_z^1 \frac{w\xi(w)}{(w^2-z^2)^{1-\mu}}\, dw. \qquad (5.32)$$

Because the solution of the simple Abel equation involves differentiation of improper integrals, the formulas given in the following exercise are frequently useful.

Exercise 5.9. Show that, for $t > 0$, differentiable $g(t)$, and $0 \leq \mu < 1$,

(a) $$\frac{d}{dt}\int_0^t \frac{g(s)}{(t-s)^\mu}\, ds = g(0)t^{-\mu} + \int_0^t \frac{g'(s)}{(t-s)^\mu}\, ds, \qquad (5.33)$$

EQUATIONS OF THE FIRST KIND

(b) $$\frac{d}{dt}\int_t^1 \frac{g(s)}{(s-t)^\mu} ds = -g(1)(1-t)^{-\mu} + \int_t^1 \frac{g'(s)}{(s-t)^\mu} ds, \qquad (5.34)$$

(c) $$\frac{d}{dt}\int_0^t \frac{sg(s)}{(t^2-s^2)^\mu} ds = g(0)t^{1-2\mu} + t\int_0^t \frac{g'(s)}{(t^2-s^2)^\mu} ds, \qquad (5.35)$$

(d) $$\frac{d}{dt}\int_t^1 \frac{sg(s)}{(s^2-t^2)^\mu} ds = -g(1)\frac{t}{(1-t^2)^\mu} + t\int_t^1 \frac{g'(s)}{(s^2-t^2)^\mu} ds. \qquad (5.36)$$

Next, we consider the generalized Abel equation

$$\int_0^t \frac{k(t,s)}{(t-s)^\mu} f(s)\, ds = g(t), \qquad (5.37)$$

under certain smoothness assumptions on $k(t,s)$ and $g(t)$.

THEOREM 5.4. *Assume that $k(t,s)$ is continuous in $0 \leq s \leq t \leq T$, and that $g(t)$ satisfies the condition (5.23) of Theorem 5.3. Then every solution of (5.37) in the class of functions $t^\beta F(t)$, with $\beta > -1$ and $F(t)$ continuous, satisfies the equation*

$$\int_0^x h(x,s)f(s)\, ds = \int_0^x \frac{g(t)}{(x-t)^{1-\mu}}\, dt, \qquad (5.38)$$

where

$$h(x,s) = \int_0^1 \frac{k(s+(x-s)u, s)}{u^\mu (1-u)^{1-\mu}}\, du. \qquad (5.39)$$

Proof. As in Theorem 5.3, we multiply (5.37) by $(x-t)^{\mu-1}$ and integrate from 0 to x; then

$$\int_0^x \int_0^t \frac{k(t,s)f(s)}{(t-s)^\mu (x-t)^{1-\mu}}\, ds\, dt = \int_0^x \frac{g(t)}{(x-t)^{1-\mu}}\, dt.$$

Interchanging order of integration using (5.22), we obtain

$$\int_0^x h(x,s)f(s)\, ds = \int_0^x \frac{g(t)}{(x-t)^{1-\mu}}\, dt,$$

where

$$h(x,s) = \int_s^x \frac{k(t,s)}{(t-s)^\mu (x-t)^{1-\mu}}\, dt.$$

The change of variables $t = s + (x-s)u$ then gives the form (5.39). □

Exercise 5.10. Show that $h(x,s)$ in (5.39) is continuous on $0 \leq s \leq x \leq T$. If $k(t,s)$ is differentiable, show that $h(x,s)$ is also differentiable.

Exercise 5.11. Show that if $k(t,t) \neq 0$ for $0 \leq t \leq T$, then $h(t,t)$ does not vanish for any t.

Exercise 5.12. Show that, if the conditions in Exercise 5.9 and Exercise 5.11 are satisfied, then (5.37) has a unique continuous solution.

To summarize, we have shown that for equations of the Abel type, analytical manipulations can be made which reduce the equations to simpler form. In the case of the simple Abel equation, closed form solutions such as (5.24) can be found. For the generalized Abel equation, no explicit inversion is known, but a reduction to an equation of the first kind with smooth kernel can be made, as shown in Theorem 5.4. Unfortunately, the integrals involved are complicated and can usually not be evaluated analytically. Consequently, numerical methods are usually required. These will be discussed in Chapter 10.

We also note, from (5.24) and (5.33) that the solution of (5.14) depends on $g'(t)$ so that Abel's equation is not well-posed. This feature plays a significant role in numerical methods.

Notes on Chapter 5. Most texts on integral equations dismiss Volterra equations of the first kind with smooth kernels with the comment that they can be reduced to equations of the second kind by differentiation. This is essentially the only technique for studying the properties of equations of the first kind.

Abel equations have received a little more attention. Tricomi [229] and Kowalewski [155] give brief accounts similar to the present chapter. A more thorough study, including smoothness results, of the general Abel equation is in [14].

Formulas for integrals involving factors of the form $(t-s)^{-\mu}$ and $(t^2-s^2)^{-\mu}$ can be found in Gradshteyn and Ryzhik [121].

Chapter 6

Convolution Equations

The applications in Chapter 2 show that many integral equations arising in practice have difference kernels of the form

$$k(t, s) = k(t-s). \tag{6.1}$$

Equations with difference kernels are referred to as convolution equations.

The linear convolution equation of the second kind is

$$f(t) = g(t) + \int_0^t k(t-s)f(s)\,ds. \tag{6.2}$$

Nonlinear convolution equations, such as

$$f(t) = g(t) + \int_0^t k(t-s)H(s, f(s))\,ds \tag{6.3}$$

are also of interest, but they will not be considered here. Similar definitions hold for convolution equations of the first kind and integrodifferential equations.

Because of the simplicity of the kernel, there exist many special results for convolution equations which cannot be extended to the general case. In this chapter we study some of the techniques for investigating the solution of convolution equations.

6.1. Some simple kernels. The situation is particularly simple when $k(t-s)$ is a polynomial in $(t-s)$. The reduction to a differential equation is then straightforward and often yields an elementary solution.

Example 6.1. Find a solution of

$$f(t) = 1 - \int_0^t (t-s)f(s)\,ds. \tag{6.4}$$

It is an elementary observation (see Theorem 3.6) that this equation has a twice continuously differentiable solution. Hence

$$f'(t) = -\int_0^t f(s) \, ds \tag{6.5}$$

and

$$f''(t) = -f(t). \tag{6.6}$$

From (6.4) and (6.5) it follows that

$$f(0) = 1, \quad f'(0) = 0,$$

so the solution of (6.6) is

$$f(t) = \cos t,$$

which indeed satisfies (6.4).

A generalization of this example is immediate. There is a close relation between differential operators and Volterra operators with kernels of the form $(t-s)^p$. For example,

$$\frac{d^2}{dt^2} \int_0^t (t-s) u(s) \, ds = u(t), \tag{6.7}$$

indicating that d^2/dt^2 is the left inverse of the Volterra operator with kernel $(t-s)$. More generally

$$\frac{d^q}{dt^q} \int_0^t (t-s)^p u(s) \, ds = p(p-1) \cdots (p-q+1) \int_0^t (t-s)^{p-q} u(s) \, ds, \tag{6.8}$$

for $p \geq 1$, $q \leq p$.

Exercise 6.1. Prove (6.8).

Exercise 6.2. Show that

$$\frac{1}{p!} \frac{d^{p+1}}{dt^{p+1}} \int_0^t (t-s)^p u(s) \, ds = u(t), \quad p \geq 0. \tag{6.9}$$

Consider now (6.2) with

$$k(t-s) = \sum_{i=0}^n a_i (t-s)^i, \tag{6.10}$$

that is, the equation

$$f(t) = g(t) + \sum_{i=0}^n a_i \int_0^t (t-s)^i f(s) \, ds. \tag{6.11}$$

If we assume that $g(t)$ is sufficiently differentiable, we get from (6.11), using

(6.8) and (6.9),

$$f'(t) = g'(t) + a_0 f(t) + \sum_{i=1}^{n} i a_i \int_0^t (t-s)^{i-1} f(s) \, ds, \tag{6.12}$$

$$f''(t) = g''(t) + a_0 f'(t) + a_1 f(t) + \sum_{i=2}^{n} i(i-1) a_i \int_0^t (t-s)^{i-2} f(s) \, ds, \tag{6.13}$$

$$\vdots$$

$$f^{(n)}(t) = g^{(n)}(t) + a_0 f^{(n-1)}(t) + a_1 f^{(n-2)}(t) \, dt + \cdots + n! \, a_n \int_0^t f(s) \, ds. \tag{6.14}$$

Differentiating once more gives the $(n+1)$st order ordinary differential equation with constant coefficients

$$f^{(n+1)}(t) = g^{(n+1)}(t) + a_0 f^{(n)}(t) + \cdots + n! \, a_n f(t). \tag{6.15}$$

From (6.11) we get

$$f(0) = g(0),$$

while (6.12)–(6.14) show that

$$f'(0) = g'(0) + a_0 f(0),$$
$$f''(0) = g''(0) + a_0 f'(0) + a_1 f(0),$$
$$f'''(0) = g'''(0) + a_0 f''(0) + a_1 f'(0) + 2 a_2 f(0),$$
$$\vdots$$

If $g^{(n+1)}$ is sufficiently simple so that a particular solution to (6.15) can be found, then a complete solution to (6.11) can be obtained using elementary results on differential equations with constant coefficients.

Exercise 6.3. Find the solution of the equation

$$f(t) + \int_0^t \{(t-s)^2 - 1\} f(s) \, ds = 1.$$

Since a kernel of the form (6.10) is degenerate, one can also use Theorem 1.1 to convert (6.11) into a system of differential equations. After simplifying a little we get the following result.

Theorem 6.1. *If $g(t)$ is continuous on $[0, T]$, then the solution to (6.11) is given by*

$$f(t) = g(t) + \sum_{i=0}^{n} a_i y_i(t), \tag{6.16}$$

where the y_i are the solution of the system

$$y_0'(t) = g(t) + \sum_{i=0}^{n} a_i y_i(t), \tag{6.17}$$

$$y_i'(t) = i y_{i-1}(t), \qquad i = 1, 2, \ldots, n, \tag{6.18}$$

$$y_0(0) = y_1(0) = \cdots = y_n(0) = 0. \tag{6.19}$$

Proof. Theorem 3.1 guarantees that (6.11) has a unique continuous solution $f(t)$. Define $z_i(t)$ by

$$z_i(t) = \int_0^t (t-s)^i f(s)\, ds, \qquad i = 0, 1, \ldots, n. \tag{6.20}$$

Then

$$g(t) + \sum_{i=0}^{n} a_i z_i(t) = g(t) + \sum_{i=0}^{n} a_i \int_0^t (t-s)^i f(s)\, ds = f(t),$$

so that the solution of (6.11) can also be written as

$$f(t) = g(t) + \sum_{i=0}^{n} a_i z_i(t). \tag{6.21}$$

Differentiating (6.20) gives

$$z_0'(t) = f(t) = g(t) + \sum_{i=0}^{n} a_i z_i(t), \tag{6.22}$$

$$z_i'(t) = \int_0^t i(t-s)^{i-1} f(s)\, ds = i z_{i-1}(t), \qquad i = 1, 2, \ldots, n. \tag{6.23}$$

Also from (6.20)

$$z_0(0) = z_1(0) = \cdots = z_n(0) = 0. \tag{6.24}$$

Thus the $z_i(t)$ satisfy (6.17) and (6.18) as well as the initial conditions (6.19) and are therefore the solution to that system. Equation (6.21) is then the desired result (6.16). □

Another equation immediately reducible to a system of differential equation is one with a kernel of the form

$$k(t-s) = \sum_{i=1}^{n} a_i e^{\beta_i(t-s)},$$

where the β_i are assumed to be all distinct.

THEOREM 6.2. *The solution of the equation*

$$f(t) = g(t) + \int_0^t \sum_{i=1}^{n} a_i e^{\beta_i(t-s)} f(s)\, ds \tag{6.25}$$

is given by

$$f(t) = g(t) - \sum_{i=1}^{n} a_i e^{\beta_i t} y_i(t), \qquad (6.26)$$

where the y_i are the solution of the system of differential equations

$$y_i'(t) = e^{-\beta_i t}\left\{g(t) - \sum_{j=1}^{n} a_j e^{\beta_j t} y_j(t)\right\}, \qquad i = 1, 2, \ldots, n, \qquad (6.27)$$

with

$$y_i(0) = 0, \qquad i = 1, 2, \ldots, n. \qquad (6.28)$$

Proof. This is an immediate consequence of Theorem 1.1 using

$$P_i(t) = a_i e^{\beta_i t}, \qquad Q_i(t) = e^{-\beta_i t}. \qquad \square$$

A closely related differential equation can be obtained with a slightly altered approach.

THEOREM 6.3. *The solution (6.25) is given by*

$$f(t) = g(t) + \sum_{i=1}^{n} F_i(t), \qquad (6.29)$$

where the F_i satisfy the system

$$F_i'(t) = a_i\left\{g(t) + \sum_{j=1}^{n} F_j(t)\right\} + \beta_i F_i(t), \qquad i = 1, 2, \ldots, n \qquad (6.30)$$

and $F_i(0) = 0$.

Proof. Let

$$F_i(t) = \int_0^t a_i e^{\beta_i(t-s)} f(s) \, ds.$$

Then

$$F_i'(t) = a_i f(t) + \beta_i \int_0^t a_i e^{\beta_i(t-s)} f(s) \, ds = a_i f(t) + \beta_i F_i(t).$$

But from (6.25)

$$f(t) = g(t) + \sum_{i=1}^{n} F_i(t),$$

so that (6.30) follows. $\square$

Obtaining a single differential equation with constant coefficients for (6.25) is a little more complicated. We begin by defining, as in Theorem 6.3

$$F_i(t) = a_i \int_0^t e^{\beta_i(t-s)} f(s) \, ds.$$

Then

$$f(t) = g(t) + \sum_{i=1}^{n} F_i(t),$$

$$f'(t) = g'(t) + \sum_{i=1}^{n} F'_i(t) = g'(t) + \sum_{i=1}^{n} a_i f(t) + \sum_{i=1}^{n} \beta_i F_i(t),$$

$$f''(t) = g''(t) + \sum_{i=1}^{n} a_i f'(t) + \sum_{i=1}^{n} a_i \beta_i f(t) + \sum_{i=1}^{n} \beta_i^2 F_i(t),$$

and in general, for $j = 1, 2, \ldots, n$,

$$f^{(j)}(t) = g^{(j)}(t) + \sum_{k=0}^{j-1} \sum_{i=1}^{n} a_i \beta_i^k f^{(j-k-1)}(t) + \sum_{i=1}^{n} \beta_i^j F_i(t). \tag{6.31}$$

Consider now the linear combination

$$\sum_{j=1}^{n} c_j f^{(j)}(t) = \sum_{j=1}^{n} c_j g^{(j)}(t) + \sum_{j=1}^{n} \sum_{k=0}^{j-1} \sum_{i=1}^{n} c_j a_i \beta_i^k f^{(j-k-1)}(t)$$

$$+ \sum_{j=1}^{n} \sum_{i=1}^{n} c_j \beta_i^j F_i(t), \tag{6.32}$$

and pick the c_j such that

$$\sum_{j=1}^{n} c_j \beta_i^j = 1, \qquad i = 1, 2, \ldots, n. \tag{6.33}$$

If all β_i are distinct and nonzero, then (6.33) has a solution since the matrix of the system is essentially a Vandermonde matrix. Then

$$\sum_{j=1}^{n} \sum_{i=1}^{n} c_j \beta_i^j F_i(t) = \sum_{i=1}^{n} F_i(t) \sum_{j=1}^{n} c_j \beta_i^j = \sum_{i=1}^{n} F_i(t) = f(t) - g(t).$$

Substituting this into (6.32) gives an equation of order n with constant coefficients for the solution $f(t)$

$$\sum_{j=1}^{n} c_j f^{(j)}(t) = \sum_{j=1}^{n} c_j g^{(j)}(t) + \sum_{j=1}^{n} \sum_{k=0}^{j-1} c_j v_k f^{(j-k-1)}(t) + f(t) - g(t), \tag{6.34}$$

where

$$v_k = \sum_{i=1}^{n} a_i \beta_i^k.$$

In general, a closed form solution to either (6.15) or (6.27) will be possible only if g is sufficiently simple. One result which sometimes helps in this is the observation made in Chapter 3, that for difference kernels the resolvent kernel itself is the solution of a convolution equation; in particular

$R(t)$ is given by

$$R(t) = k(t) + \int_0^t k(t-s) R(s)\, ds. \tag{6.35}$$

Suppose now that $k(t)$ is of polynomial form (6.10). The equation for $R(t)$ is then of the form (6.11) with

$$g(t) = \sum_{i=0}^n a_i t^i. \tag{6.36}$$

Combining this with (6.15), we get a simple result for the resolvent kernel.

THEOREM 6.4. *If $k(t)$ is of polynomial form (6.10), then its resolvent kernel $R(t)$ is the solution of the homogeneous linear differential equation with constant coefficients*

$$R^{(n+1)}(t) = a_0 R^{(n)}(t) + a_1 R^{(n-1)}(t) + \cdots + n!\, a_n R(t), \tag{6.37}$$

with initial conditions

$$R(0) = a_0,$$
$$R'(0) = a_1 + a_0^2,$$
$$R''(0) = 2a_2 + a_0(a_1 + a_0^2) + a_1 a_0,$$
$$\vdots$$

Proof. Use (6.15), with $g(t)$ given by (6.36). Then $g^{(n+1)}(t) = 0$. The initial conditions are those for (6.15) using the specific form of $g(t)$. □

Closely related to the resolvent kernel $R(t)$ is the so-called *differential resolvent* $S(t)$ defined as the solution of the equation

$$S(t) = 1 + \int_0^t k(t-s) S(s)\, ds. \tag{6.38}$$

By a simple change of variable, we see that $S(t)$ also satisfies

$$S(t) = 1 + \int_0^t k(\tau) S(t-\tau)\, d\tau, \tag{6.39}$$

so that

$$S'(t) = k(t) S(0) + \int_0^t k(\tau) S'(t-\tau)\, d\tau.$$

But from (6.38) it follows that $S(0) = 1$; consequently

$$S'(t) = k(t) + \int_0^t k(\tau) S'(t-\tau)\, d\tau$$
$$= k(t) + \int_0^t k(t-s) S'(s)\, ds.$$

Since this is exactly the same equation as (6.35), it follows that

$$S'(t) = R(t). \tag{6.40}$$

6.2. Laplace transforms. One of the primary tools for the solution and analysis of linear convolution equations is the Laplace transform. In this section we present, generally without proof, those aspects of Laplace transform theory most useful in the study of integral equations. Further details, as well as the proofs, can be found in books on the Laplace transform.

For convenience we will use the usual notation for the convolution of two functions

$$(f_1 * f_2)(t) = \int_0^t f_1(t-s) f_2(s) \, ds. \tag{6.41}$$

The Laplace transform of a given function $f(t)$ will be denoted by f^* or $\mathcal{L}(f)$; it is defined as

$$f^*(w) = \mathcal{L}(f)(w) = \int_0^\infty e^{-wt} f(t) \, dt, \tag{6.42}$$

where w is a complex number. This definition holds provided the integral in (6.42) is convergent, which is generally so for all w in the half-plane $\operatorname{Re}(w) \geq w_0$, for some w_0. In the half-plane of convergence $f^*(w)$ is an analytic function defined by (6.42). To the left of $\operatorname{Re}(w) = w_0$, the Laplace transform $f^*(w)$ is defined by analytic continuation. The Laplace transform operator $\mathcal{L}$ has an inverse $\mathcal{L}^{-1}$, given by

$$\mathcal{L}^{-1}(u)(t) = \frac{1}{2\pi i} \int_{a-i\infty}^{a+i\infty} e^{wt} u(w) \, dw, \tag{6.43}$$

which is such that

$$\mathcal{L}^{-1}(f^*)(t) = f(t). \tag{6.44}$$

The real number a in (6.43) has to be such that all singularities of $f^*(w)$ lie to the left of the path of integration.

The usefulness of Laplace transforms in the solution of convolution equations arises primarily from the following standard result.

THEOREM 6.5. *Let f_1 and f_2 be two functions which are absolutely integrable over some interval $[0, T]$ and which are bounded in every finite subinterval not including the origin. If, furthermore $\mathcal{L}(f_1)$ and $\mathcal{L}(f_2)$ are absolutely convergent for $w \geq w_0$, then*

$$\mathcal{L}(f_1 * f_2) = \mathcal{L}(f_1) \cdot \mathcal{L}(f_2), \qquad \operatorname{Re} w \geq w_0. \tag{6.45}$$

In other words, the Laplace transform of a convolution is the product of the individual transforms.

CONVOLUTION EQUATIONS

Proof. This is a standard result which can be found in any book on the Laplace transform, for example [239]. □

Consider now the linear convolution equation of the second kind (6.2), which can be written in shorthand notation as

$$f = g + k * f. \tag{6.46}$$

Applying the Laplace transform to this, noting that $\mathscr{L}$ is a linear operator, and using Theorem 6.5, we have

$$f^* = g^* + k^* f^*. \tag{6.47}$$

Solving this equation for f^*,

$$f^* = \frac{g^*}{1 - k^*}, \tag{6.48}$$

and applying the inverse Laplace transform, gives the solution

$$f = \mathscr{L}^{-1}\left(\frac{g^*}{1 - k^*}\right). \tag{6.49}$$

While (6.49) gives the solution of (6.46) in terms of definite integrals, its practical use for finding the solution of integral equations is limited by the fact that the inverse Laplace transform is known only for some rather simple functions. Numerically, (6.49) is also not easy to use since it involves integration in the complex plane. There are easier ways to solve Volterra equations numerically, as we shall see. The main advantage of the Laplace transform method is that it allows us to obtain some qualitative properties of the solution in a simple and elegant way, as will be demonstrated in subsequent sections of this chapter. Still, in some simple cases, (6.49) can be used to obtain a solution to (6.46).

Example 6.2. Find the solution of the equation

$$f(t) = 1 + \int_0^t \sin(t-s) f(s) \, ds.$$

From a table of Laplace transforms we find

$$\mathscr{L}(1)(w) = \frac{1}{w}, \qquad \mathscr{L}(\sin t)(w) = \frac{1}{w^2 + 1}.$$

Therefore

$$f^*(w) = \frac{1/w}{1 - 1/(w^2 + 1)} = \frac{1}{w} + \frac{1}{w^3},$$

and

$$f(t) = \mathscr{L}^{-1}\left(\frac{1}{w} + \frac{1}{w^3}\right) = \mathscr{L}^{-1}\left(\frac{1}{w}\right) + \mathscr{L}^{-1}\left(\frac{1}{w^3}\right) = 1 + \frac{t^2}{2}.$$

Exercise 6.4. Use the Laplace transform method to solve the equation
$$f(t) = 1 + \int_0^t e^{\alpha(t-s)} f(s) \, ds.$$

Exercise 6.5. Use the Laplace transform method to solve the equation of the first kind
$$\int_0^t e^{-(t-s)} f(s) \, ds = e^{-t} + t - 1.$$

6.3. Solution methods using Laplace transforms. Equation (6.49) can be used to obtain a solution only in very special cases for which $g(t)$ and $k(t)$ are quite simple. A more general expression involves only the Laplace transform of k; to derive it we use the fact that for convolution equations the resolvent satisfies a convolution equation. The resolvent kernel satisfies equation (6.35), which in convolution form is

$$R = k + k * R. \tag{6.50}$$

Taking Laplace transforms and solving for R^* gives

$$R^* = k^* + k^* R^*, \qquad R^* = \frac{k^*}{1 - k^*},$$

so that

$$R = \mathcal{L}^{-1}\left(\frac{k^*}{1 - k^*}\right). \tag{6.51}$$

Substituting this into (3.25) gives the solution

$$f = g + \mathcal{L}^{-1}\left(\frac{k^*}{1 - k^*}\right) * g. \tag{6.52}$$

In cases where g is complicated, but k is relatively simple, this can be useful.

Exercise 6.6. Show that, for arbitrary $g(t)$, the solution of
$$f(t) = g(t) + \int_0^t \sin(t - s) f(s) \, ds$$
is given by
$$f(t) = g(t) + \int_0^t (t - s) g(s) \, ds.$$

Another closed form solution useful for simple $k(t)$ and general $g(t)$ uses the differential resolvent.

THEOREM 6.6. *Let $u(t)$ be the solution of the equation*

$$u(t) = 1 + \int_0^t k(t - s) u(s) \, ds. \tag{6.53}$$

Then the solution of

$$f(t) = g(t) + \int_0^t k(t-s)f(s)\, ds \qquad (6.54)$$

is given by

$$f(t) = g(0)u(t) + \int_0^t u(t-s)g'(s)\, ds. \qquad (6.55)$$

This conclusion holds for all $k(t)$ and $g(t)$ for which the formal manipulations in the following argument are justified.

Proof. Take Laplace transforms of (6.53) and (6.54); this gives

$$u^*(w) = \frac{1}{w\{1 - k^*(w)\}}$$

and

$$f^*(w) = \frac{g^*(w)}{1 - k^*(w)}.$$

Thus

$$\frac{f^*(w)}{u^*(w)} = wg^*(w). \qquad (6.56)$$

We now use another elementary result on Laplace transforms, namely that, for sufficiently smooth functions y,

$$\mathscr{L}(y^{(p)})(w) = w^p y^*(w) - \sum_{i=0}^{p-1} y^{(i)}(0) w^{p-i-1}. \qquad (6.57)$$

Applying this to (6.56) with $y = g$ and $p = 1$, we find that

$$\frac{f^*}{u^*} = g(0) + \mathscr{L}(g'),$$

or

$$\mathscr{L}(f) = \mathscr{L}(u)g(0) + \mathscr{L}(u)\mathscr{L}(g').$$

By Theorem 6.5 then

$$\mathscr{L}(f) = \mathscr{L}(u)g(0) + \mathscr{L}(u * g'),$$

and applying $\mathscr{L}^{-1}$,

$$f = ug(0) + u * g',$$

which is the desired result (6.55). $\square$

As indicated by Exercise 6.5, the Laplace transform method can also be used on equations of the first kind. Again, some manipulations are required to find general forms. We illustrate this by deriving the solution of the simple Abel equation by this technique.

Consider the equation

$$\int_0^t \frac{f(s)}{\sqrt{t-s}}\, ds = g(t).$$

Note that the kernel, although not bounded, satisfies the conditions of Theorem 6.5. Applying $\mathscr{L}$ to this equation, and noting that

$$\mathscr{L}\left(\frac{1}{\sqrt{t}}\right)(w) = \frac{\sqrt{\pi}}{\sqrt{w}}, \tag{6.58}$$

we obtain

$$f^*(w) = \frac{\sqrt{w}}{\sqrt{\pi}} g^*(w). \tag{6.59}$$

An immediate application of $\mathscr{L}^{-1}$ does not yield anything very useful; instead we introduce a new variable $z(t)$ such that

$$f(t) = z'(s).$$

Then, from (6.57),

$$f^*(w) = wz^*(w) - z(0),$$

and from (6.59)

$$z^*(w) = \frac{z(0)}{w} + \frac{1}{\sqrt{\pi}} \frac{1}{\sqrt{w}} g^*(w).$$

From (6.58) this is the same as

$$\mathscr{L}(z) = \frac{z(0)}{w} + \frac{1}{\pi} \mathscr{L}\left(\frac{1}{\sqrt{t}}\right) \mathscr{L}(g),$$

and by Theorem 6.5

$$\mathscr{L}(z) = \frac{z(0)}{w} + \frac{1}{\pi} \mathscr{L}\left(\frac{1}{\sqrt{t}} * g\right).$$

This can now be inverted to give

$$z(t) = z(0) + \frac{1}{\pi} \int_0^t \frac{g(s)}{\sqrt{t-s}} ds. \tag{6.60}$$

Finally, differentiating (6.60), we arrive, as expected, at

$$f(t) = \frac{1}{\pi} \frac{d}{dt} \int_0^t \frac{g(s)}{\sqrt{t-s}} ds. \tag{6.61}$$

As a final example, consider the integrodifferential equation with constant coefficients of the form

$$\sum_{j=0}^n a_j \frac{d^j}{dt^j} f(t) = g(t) + \int_0^t k(t-s) f(s) \, ds, \tag{6.62}$$

with

$$f(0) = f'(0) = \cdots = f^{(n-1)}(0) = 0.$$

Applying $\mathscr{L}$ to (6.62) and using (6.57), we obtain

$$\sum_{j=0}^{n} a_j w^j f^*(w) = g^*(w) + k^*(w) f^*(w),$$

so that the solution of (6.62) is

$$f = \mathscr{L}^{-1}\left\{\frac{g^*(w)}{\sum_{j=0}^{n} a_j w^j - k^*(w)}\right\}. \tag{6.63}$$

The manipulations in these examples are purely formal and one needs to demonstrate their validity by showing that all necessary conditions are satisfied. Such a priori justification then requires some analysis of the solution of the integral equations using the results of Chapter 3. Perhaps a more practical procedure is to use the Laplace transform method to arrive at a possible solution, then substitute it into the original equation to verify it.

While the Laplace transform method can give explicit solutions for special integral equations, its usefulness becomes more apparent when we consider some qualitative properties of the solution.

6.4. The asymptotic behavior of the solution for some special equations.
The results developed in Chapter 3 can of course be applied to convolution equations to get some elementary results. For example, Theorem 3.8 becomes

THEOREM 6.7. *If the kernel $k(t)$ in the equation*

$$f(t) = 1 + \int_0^t k(t-s) f(s)\, ds \tag{6.64}$$

satisfies

$$k(t) < 0, \tag{6.65}$$

$$k'(t) > 0, \tag{6.66}$$

then the solution of (6.64) satisfies

$$0 \leq f(t) \leq 1. \tag{6.67}$$

Proof. This is an immediate consequence of Theorem 3.8. ☐

To obtain more far-reaching results the Laplace transform method is often useful. We will consider there only some linear convolution equation with certain special, but important, forms.

First, consider

$$f(t) = k(t) + \int_0^t k(t-s) f(s)\, ds. \tag{6.68}$$

This equation is of some significance. As already pointed out, its solution is the resolvent kernel $R(t)$. Also, it arises in practice as the renewal equation discussed in Chapter 2. Consequently there is considerable interest in this form.

If (6.68) is a renewal equation, then $k(t)$ represents a probability density and must satisfy

$$\int_0^\infty k(s)\,ds = 1, \tag{6.69}$$

$$k(t) \geq 0. \tag{6.70}$$

We will therefore consider (6.68), subject to conditions (6.69) and (6.70).

Taking Laplace transforms of (6.68) gives

$$f^*(w) = \frac{k^*(w)}{1 - k^*(w)}. \tag{6.71}$$

Let us assume now that the right-hand side of (6.71) can be written as a rational function of w, that is,

$$\frac{k^*(w)}{1 - k^*(w)} = \frac{1 + a_1 w + a_2 w^2 + \cdots + a_n w^n}{b_0 + b_1 w + \cdots + b_m w^m}. \tag{6.72}$$

In many cases of interest this condition is fulfilled.

If (6.72) holds, we factor the denominator to give

$$\frac{k^*(w)}{1 - k^*(w)} = \frac{1 + a_1 w + a_2 w^2 + \cdots + a_n w^n}{(w - \beta_0)(w - \beta_1) \cdots (w - \beta_m)}. \tag{6.73}$$

Let us ignore the possibility of multiple roots (they introduce no essential complications) and order the roots so that

$$\mathrm{Re}\,(\beta_m) \leq \mathrm{Re}\,(\beta_{m-1}) \leq \cdots \leq \mathrm{Re}\,(\beta_0).$$

From the definition of the Laplace transform and from (6.69) and (6.70) it follows that $k^*(0) = 1$ and that $k^*(w) < 1$ for all w with $\mathrm{Re}\,(w) > 0$, or $\mathrm{Re}\,(w) = 0$ and $\mathrm{Im}\,(w) \neq 0$. From this we conclude that $\beta_0 = 0$ and that $\mathrm{Re}\,(\beta_i) < 0$ for $i = 1, 2, \ldots, m$.

If we use a partial fraction decomposition to write

$$\frac{k^*(w)}{1 - k^*(w)} = \frac{c_0}{w} + \frac{c_2}{(w - \beta_1)} + \cdots + \frac{c_m}{(w - \beta_m)}, \tag{6.74}$$

then c_0 can be obtained by

$$c_0 = \lim_{w \to 0} \frac{w k^*(w)}{1 - k^*(w)}. \tag{6.75}$$

If we now apply $\mathscr{L}^{-1}$ to (6.71) and use (6.74), we get

$$f(t) = \mathscr{L}^{-1}\left(\frac{k^*}{1-k^*}\right)$$

$$= c_0 \mathscr{L}^{-1}\left(\frac{1}{w}\right) + \sum_{i=1}^{m} c_i \mathscr{L}^{-1}\left(\frac{1}{w-\beta_i}\right)$$

$$= c_0 + \sum_{i=1}^{m} c_i e^{\beta_i t}.$$

Since $\mathrm{Re}\,(\beta_i) < 0$ for $i = 1, 2, \ldots, m$ it follows that

$$\lim_{t \to \infty} f(t) = c_0. \tag{6.76}$$

Also, because

$$1 - k^*(w) = \int_0^\infty (1 - e^{-wt}) k(t) \, dt,$$

it follows that

$$\lim_{w \to 0} \frac{1}{w} (1 - k^*(w)) = \int_0^\infty t k(t) \, dt.$$

The conclusion then is that, if $f(t)$ is the solution of (6.68) subject to conditions (6.69), (6.70), and (6.72), then

$$\lim_{t \to \infty} f(t) = \frac{1}{\int_0^\infty t k(t) \, dt}. \tag{6.77}$$

More precise and more extensive results can be obtained using various known properties of the Laplace transform. A set of closely related results, the so-called *Tauberian theorems*, are particularly useful. One of the simplest of these is the following.

THEOREM 6.8. *If $f(t) \geq 0$ and*

$$\lim_{w \to 0^+} w^p \mathscr{L}(f)(w) = c, \tag{6.78}$$

with positive c and p, then, for large t

$$\int_0^t f(t) \, dt = \frac{c t^p}{\Gamma(p+1)} (1 + \varepsilon(t)), \tag{6.79}$$

where $\varepsilon(t) \to 0$ as $t \to \infty$.

Proof. See Bellman and Cooke [26, p. 240]. □

To see the use of this theorem, consider the equation

$$f(t) = 1 + \int_0^t k(t-s) f(s) \, ds, \tag{6.80}$$

subject to
$$\int_0^\infty k(t)\,dt = 1, \qquad (6.81)$$

$$\int_0^\infty tk(t)\,dt = m < \infty, \qquad (6.82)$$

$$k(t) \geq 0. \qquad (6.83)$$

Applying $\mathscr{L}$ to (6.80) gives
$$f^*(w) = \frac{1}{w[1 - k^*(w)]}. \qquad (6.84)$$

Now
$$k^*(w) = k^*(0) + w\frac{d}{dw}k^*(0) + \cdots,$$

so that
$$k^*(w) = 1 - mw + \cdots.$$

Therefore $f^*(w)$ satisfies the conditions of Theorem 6.8 with $c = 1/m$, $p = 2$ and we can conclude that
$$\int_0^t f(s)\,ds = \frac{t^2}{2m}(1 + \varepsilon(t)), \qquad (6.85)$$

where $\lim_{t\to\infty} \varepsilon(t) = 0$.

Exercise 6.7. If $f(t)$ is the solution of
$$f(t) = t^p + \int_0^t k(t-s)f(s)\,ds,$$

where p is a positive integer and $k(t)$ satisfies conditions (6.81)–(6.83), show that
$$\int_0^t f(s)\,ds = O(t^{p+2})$$

for large t.

Notes on Chapter 6. The reduction of convolution equations to systems of differential equations, and consequent use in obtaining approximate solutions, is studied in [33].

The results on Laplace transforms quoted in this chapter can be found in standard texts, such as Widder [239]. Various Tauberian theorems, including Theorem 6.8, are given in Bellman and Cooke [26].

Because of their importance in applications, convolution equations have received a great deal of attention. The books by Bellman [25] and Bellman and Cooke [26] contain a wealth of material on the linear case. Nonlinear convolution equations have also been studied extensively. Some representative papers on this subject are [40], [66], [105], [162], and [234].

PART B: NUMERICAL METHODS

Chapter 7

The Numerical Solution of Equations of the Second Kind

Since few of the Volterra equations encountered in practice can be solved explicitly, it is often necessary to resort to numerical techniques. There are many alternatives available as we shall see; we will concentrate here on the underlying ideas on which these methods are based. In order to bring out the essential ideas clearly and to avoid unnecessary complications, we use the simplest setting and consider the equation

$$f(t) = g(t) + \int_0^t K(t, s, f(s))\, ds, \qquad 0 \le t \le T, \qquad (7.1)$$

under the conditions
   (i) $g(t)$ is a continuous function in $0 \le t \le T$,
   (ii) the kernel $K(t, s, y)$ is continuous in $0 \le s \le t \le T$, $-\infty < y < \infty$,
   (iii) the kernel satisfies the Lipschitz condition

$$|K(t, s, y_1) - K(t, s, y_2)| \le L|y_1 - y_2|, \qquad (7.2)$$

for all $0 \le s \le t \le T$, and all y_1, y_2.

As shown in Chapter 4, these conditions are sufficient to guarantee that (7.1) has a unique continuous solution. The analysis of the numerical methods will utilize these assumptions, and, strictly speaking, holds only when they are satisfied. The algorithms can usually be applied to other equations as well, although not all conclusions are necessarily valid. For certain types of kernels it may actually be necessary to modify the procedures, as we shall see in the next chapter.

We begin our discussion in this chapter by presenting a rather simple and intuitively reasonable method based on the trapezoidal integration rule. Methods based on more accurate integration methods are investigated next. The results suggest the possibility of using various standard numerical integration techniques for the approximate solution of Volterra equations.

To justify this, a general convergence theorem is presented in § 7.3. In §§ 7.4 and 7.5 we investigate the question of error estimates and numerical stability. A class of methods, closely related to the implicit Runge–Kutta methods for ordinary differential equations, is presented in § 7.6. A detailed numerical example in § 7.7 gives some practical evidence for the effectiveness of these methods. In § 7.8 we show how explicit Runge–Kutta methods, popular for the solution of ordinary differential equations, can be generalized to Volterra equations. Finally, in § 7.9 we present a short summary of various ideas that have been proposed but not treated in detail here.

7.1. A simple numerical procedure. Intuitively, we can compute an approximate solution to (7.1) by the following simple process. Suppose that for a given stepsize $h > 0$ we know the solution at points $t_i = ih$, $i = 0, 1, \ldots, n-1$. An approximation to $f(t_n)$ can then be computed by replacing the integral on the right side of (7.1) by a numerical integration rule using values of the integrand at t_i, $i = 0, 1, \ldots, n$ and solving the resulting equation for $f(t_n)$. Since $f(t_0) = g(0)$, the approximate solution can be computed in this step-by-step fashion. The particulars of the algorithm depend primarily on the integration rule we choose; to start let us consider the very simple procedure one obtains by the use of the (composite) *trapezoidal* rule.

If we let F_n denote the approximate value of $f(t_n)$, we can compute F_n by

$$F_n = g(t_n) + h\left\{\tfrac{1}{2}K(t_n, t_0, F_0) + \sum_{i=1}^{n-1} K(t_n, t_i, F_i) + \tfrac{1}{2}K(t_n, t_n, F_n)\right\},$$

$$n = 1, 2, 3, \ldots \quad (7.3)$$

with $F_0 = g(0)$.

The unknown F_n is defined by (7.3) implicitly, but for sufficiently small h the equation has a unique solution. In the linear case we can of course solve it directly for F_n; in the nonlinear case we would normally use some iterative technique to solve for F_n to within a desired accuracy.

A formal error analysis of this procedure will be given later. For the moment let us proceed in a more empirical fashion and see what results we obtain for a simple test example.

Example 7.1. The equation

$$f(t) = e^t - \int_0^t e^{(t-s)} f(s)\, ds \qquad (7.4)$$

has exact solution

$$f(t) = 1. \qquad (7.5)$$

The errors in the approximate solution by the trapezoidal method are shown in Table 7.1 for several stepsizes.

TABLE 7.1
Observed errors $F_i - f(t_i)$ for Example 7.1.

t	$h = 0.1$	$h = 0.05$	$h = 0.025$
0.1	-8.3×10^{-5}	-2.1×10^{-5}	-5.2×10^{-6}
0.2	-1.7×10^{-4}	-4.2×10^{-5}	-1.0×10^{-5}
0.3	-2.5×10^{-4}	-6.3×10^{-5}	-1.6×10^{-5}
0.4	-3.3×10^{-4}	-8.3×10^{-5}	-2.1×10^{-5}
0.5	-4.2×10^{-4}	-1.0×10^{-4}	-2.6×10^{-5}
0.6	-5.0×10^{-4}	-1.3×10^{-4}	-3.1×10^{-5}
0.7	-5.8×10^{-4}	-1.5×10^{-4}	-3.6×10^{-5}
0.8	-6.7×10^{-4}	-1.7×10^{-4}	-4.2×10^{-5}
0.9	-7.5×10^{-4}	-1.9×10^{-4}	-4.7×10^{-5}
1.0	-8.3×10^{-4}	-2.1×10^{-4}	-5.2×10^{-5}

From these results we can observe that the method apparently converges, that is, the approximate solution becomes more accurate as h decreases. Furthermore, we note that the error is approximately proportional to h^2. This $O(h^2)$ dependence of the error is a reflection of the second order convergence of the trapezoidal method. This leads us to suspect that, in general, if the integral in (7.1) is approximated by a numerical integration having a certain order of accuracy, then the approximate solution computed this way has the same order of accuracy. We shall see shortly that this conjecture is essentially true.

Exercise 7.1. Use the trapezoidal method with $h = 0.1$, 0.05, 0.025 to compute an approximate solution to

$$f(t) = 1 + \int_0^t \sin(t-s)f(s)\, ds.$$

Compare with the exact solution to show the $O(h^2)$ dependence of the error.

7.2. Methods based on more accurate numerical integration. The trapezoidal method is quite simple but the results are of relatively low accuracy. If the conjecture made in the previous section is true and the accuracy of the approximate solution depends on the accuracy of the numerical integration, then for more accurate methods better integration rules must be used. Some restrictions and practical complications arise immediately. Because of the stepwise way of computing the solution, we must rely largely on integration rules with constant stepsize.[1] We assume therefore that we have an integration rule of the form

$$\int_0^{nh} \varphi(t)\,dt \simeq h \sum_{i=0}^{n} w_{ni} \varphi(t_i), \qquad (7.6)$$

where $\varphi(t)$ is any continuous integrand. The w_{ni} are called the *integration weights*. Here h denotes the constant stepsize for the integration, and $t_i = ih$.

Using this to replace the integral in (7.1), we are led to consider the numerical method

$$F_n = g(t_n) + h \sum_{i=0}^{n} w_{ni} K(t_n, t_i, F_i), \qquad n = r, r+1, \ldots. \qquad (7.7)$$

If, as is generally the case, the weights w_{ni} are uniformly bounded, then the equation has a unique solution for all sufficiently small h.

First we note that (7.7) holds only for $n \geq r$, where r is some fixed integer. This reflects the fact that higher order integration rules require a minimum number of points. The values of $F_1, F_2, \ldots, F_{r-1}$ can therefore not be obtained by (7.7) and must be computed some other way. How these *starting values* can be found will be discussed later, but apart from this (7.7) represents a viable computational procedure.

Within the restriction of constant stepsize we have the choice of using either *Newton-Cotes* or *Gregory* type integration rules. For example, the fourth order Gregory formula has weights (for $n \geq 5$)

$$w_{n0} = w_{nn} = \tfrac{3}{8},$$
$$w_{n1} = w_{n,n-1} = \tfrac{7}{6},$$
$$w_{n2} = w_{n,n-2} = \tfrac{23}{24},$$
$$w_{ni} = 1, \qquad i = 3, 4, \ldots, n-3. \qquad (7.8)$$

Thus (7.7) with these weights and $r = 5$ will be called the *fourth order Gregory method* for solving the Volterra equation. The value $r = 5$ arises because at least six points are required for (7.8) to hold. The values of F_1, F_2, F_3, F_4 need to be given to start; F_0 is of course always taken as $f(0) = g(0)$.

Exercise 7.2. Find the weights for the sixth order Gregory formula. How many starting values are needed in (7.7)?

---

[1] It is of course possible to vary the stepsize during the computation, for example, by some adaptive approach. However, this does not affect the subsequent discussion in any significant way.

Any of the Gregory formulas can be used, giving a class of numerical methods for solving (7.1).

The use of the Newton–Cotes type formulas introduces a slight complication, since these rules involve some restriction on the number of points. For example, *Simpson's rule* (that is, the repeated Simpson's rule) can be applied only when n is even. For odd n some adjustment has to be made. One way is to apply the so-called *three-eighths rule* over four adjacent points and the Simpson's rule over the rest of the interval. If the three-eighths rule is used on the points t_0, t_1, t_2, t_3, one gets the weights (for $n \geq 2$)

n is even: $\quad w_{n0} = w_{nn} = \frac{1}{3},$

$\qquad\qquad\quad w_{n,2i} = \frac{2}{3}, \qquad i = 0, 1, \ldots, n/2 - 1,$

$\qquad\qquad\quad w_{n,2i+1} = \frac{4}{3}, \qquad i = 0, 1, \ldots, n/2 - 1,$

n is odd: $\quad w_{n0} = \frac{3}{8},$

$\qquad\qquad\quad w_{n1} = w_{n2} = \frac{9}{8},$

$\qquad\qquad\quad w_{n3} = \frac{17}{24} - \frac{1}{3}\delta_{n3},$

$\qquad\qquad\quad w_{n,2i} = \frac{4}{3}, \qquad i = 2, 3, \ldots, (n-1)/2,$

$\qquad\qquad\quad w_{n,2i+1} = \frac{2}{3}, \qquad i = 2, 3, \ldots, (n-3)/2,$

$\qquad\qquad\quad w_{nn} = \frac{1}{3}, \qquad n \geq 5.$

We will call this *Simpson's method* 1.

If we use the three-eighths rule at the upper end, that is, on the points t_{n-3}, t_{n-2}, t_{n-1}, t_n, we get the weights

n is even: $\quad$ as in Simpson's method 1,

n is odd: $\quad w_{n0} = \frac{1}{3}, \qquad\qquad n \geq 5,$

$\qquad\qquad\quad w_{n,2i} = \frac{2}{3}, \qquad i = 1, 2, \ldots, (n-5)/2,$

$\qquad\qquad\quad w_{n,2i+1} = \frac{4}{3}, \qquad i = 0, 1, \ldots, (n-5)/2,$

$\qquad\qquad\quad w_{n,n-3} = \frac{17}{24} - \frac{1}{3}\delta_{n3},$

$\qquad\qquad\quad w_{n,n-1} = w_{n,n-2} = \frac{9}{8},$

$\qquad\qquad\quad w_{nn} = \frac{3}{8}.$

In these formulas, as well as in subsequent discussions, δ_{ij} denotes the Kronecker delta

$$\delta_{ij} = 0, \quad i \neq j, \qquad \delta_{ii} = 1.$$

We will refer to this second form as *Simpson's method* 2. Both forms of Simpson's method require starting values F_0 and F_1. While intuitively there seems to be little difference between the two methods, we shall see later that Simpson's method 2 is preferable for computational purposes.

Exercise 7.3. Derive the weights for both forms of Simpson's method.

Methods based on higher order Newton–Cotes formulas can be constructed in a similar way, but it becomes necessary to combine several different rules. The resulting complication limits the usefulness of this type of method.

Exercise 7.4. Investigate the construction of a method of the form (7.7) based on Newton–Cotes formulas of order at least five.

This intuitive approach leads us to a number of different plausible methods for the approximate solution of (7.1). To justify them and to predict their relative usefulness requires a detailed error analysis.

7.3. Error analysis: convergence of the approximate solution. The methods discussed in the previous section determine an approximate solution only at the error at the points t_i; thus we can talk only at the error at these points. In practice this is usually adequate. If it is necessary to produce an approximate solution in the form of a continuous function, one has several choices. One can either construct an interpolating polynomial using the computed values of the solution at t_i or one can use special methods yielding continuous functions directly. Algorithms for this latter approach differ somewhat in the practical details from the methods discussed here, but the underlying principles are very much the same.

Let us therefore consider the set of values

$$\varepsilon_i = F_i - f(t_i), \qquad i = 0, 1, 2, \ldots \tag{7.9}$$

which we will call the *discretization error*. We are interested in the behavior of this discretization error as a function of the stepsize h. For any particular value of h the interval $[0, T]$ is divided into N parts in such a way that $Nh = T$. Therefore, as $h \to 0$, N goes to infinity in such a way that $Nh = T$. This implies that h can take on only certain *admissible* values, namely such that $h \leq T$ and T/h is an integer. Since this particular way of approaching the limit occurs frequently in subsequent discussions, we will use for simplicity the notation $\lim_h$ for it.

DEFINITION 7.1. A method of the form (7.7) is said to be a *convergent approximation method* (for some equation or class of equations) if

$$\lim_h \left(\max_{0 \leq i \leq N} |\varepsilon_i| \right) = 0. \tag{7.10}$$

DEFINITION 7.2. If, for all admissible h, there exists a number $M < \infty$, independent of h, such that

$$\max_{0 \leq i \leq N} |\varepsilon_i| \leq M h^p, \tag{7.11}$$

and if p is the largest number for which such an inequality holds, then p is called the *order of convergence* of the method.

The order of convergence is, not surprisingly, closely connected with the accuracy of the numerical integration employed.

DEFINITION 7.3. Let f be the solution of (7.1). Then the function

$$\delta(h, t_n) = \int_0^{t_n} K(t_n, s, f(s))\, ds - h \sum_{i=0}^{n} w_{ni} K(t_n, t_i, f(t_i)) \qquad (7.12)$$

is the *local consistency error* for (7.1).

The local consistency error is a measure of the accuracy with which, in the context of a given equation, the numerical integration rule represents the integral.

DEFINITION 7.4. Let $\mathscr{C}$ be a class of equations of the form (7.1). If for every equation in $\mathscr{C}$

$$\lim_h \max_{0 \leq n \leq N} |\delta(h, t_n)| = 0, \qquad (7.13)$$

then the approximation method (7.7) is said to be *consistent* with (7.1) for the class of equations $\mathscr{C}$. If for every equation in $\mathscr{C}$, there exists a constant c (independent of h, but generally dependent on K and f) such that

$$\max_{0 \leq n \leq N} |\delta(h, t_n)| \leq ch^p, \qquad (7.14)$$

then the method is said to be *consistent of order p* in $\mathscr{C}$.

Before proceeding with the statement and proof of a convergence theorem, we need the following basic result.

THEOREM 7.1. *Let the sequence $\xi_0, \xi_1, \ldots$ satisfy*

$$|\xi_n| \leq A \sum_{i=0}^{n-1} |\xi_i| + B_n, \qquad n = r, r+1, \ldots, \qquad (7.15)$$

where

$$A > 0, \qquad |B_n| \leq B, \qquad \sum_{i=0}^{r-1} |\xi_i| \leq \eta. \qquad (7.16)$$

Then

$$|\xi_n| \leq (1+A)^{n-r}(B + A\eta), \qquad n = r, r+1, \ldots. \qquad (7.17)$$

Proof. The result is easily established with an inductive argument. We omit the details. □

Exercise 7.5. Prove Theorem 7.1.

From (7.17) it follows that if $A = hK$, $t_n = nh$, then

$$|\xi_n| \leq (B + hK\eta)e^{Kt_n}. \qquad (7.18)$$

The main convergence theorem follows by a simple application of this preliminary result.

THEOREM 7.2. *Consider the approximate solution of* (7.1) *by* (7.7) *and assume that*

(i) *the solution $f(t)$ of* (7.1) *and the kernel $K(s, t, f)$ are such that the approximation method is consistent of order p with* (7.1),

(ii) *the weights satisfy*

$$\sup_{n, i} |w_{ni}| \leq W < \infty,$$

(iii) *the starting errors $F_i - f(t_i)$, $i = 0, 1, \ldots, r-1$ go to zero as $h \to 0$. Since r is fixed, this implies that*

$$\lim_{h \to 0} \sum_{i=0}^{r-1} |F_i - f(t_i)| = 0.$$

Then the method is a convergent approximation method. Also, in the absence of starting errors, the order of convergence is at least p.

Proof. Putting $t = t_n$ in (7.1) and subtracting from (7.7), we get for $n = r, r+1, \ldots$

$$\varepsilon_n = h \sum_{i=0}^{n} w_{ni} \{K(t_n, t_i, F_i) - K(t_n, t_i, f(t_i))\} - \delta(h, t_n).$$

Using the Lipschitz conditions (7.2) and assumption (ii), and choosing $h < 1/LW$, we have

$$|\varepsilon_n| \leq \frac{hWL}{1 - hWL} \sum_{i=0}^{n-1} |\varepsilon_i| + \frac{|\delta(h, t_n)|}{1 - hWL}.$$

Applying Theorem 7.1 and (7.18) gives

$$|\varepsilon_n| \leq \frac{1}{1 - hWL} \left\{ \max_{r \leq i \leq n} |\delta(h, t_i)| + hWL \sum_{i=0}^{r-1} |F_i - f(t_i)| \right\} e^{WLt_n/1 - hWL}$$

(7.19)

Since by assumption both the starting errors and the local consistency error are zero in the limit, it follows that

$$\lim_h |\varepsilon_n| = 0.$$

If there are no starting errors, then

$$|\varepsilon_n| = O(\max |\delta(h, t_i)|),$$

and the proof is complete. □

One further conclusion can be drawn from (7.19): the effect of the starting errors is attenuated by a factor of h, showing that these methods are relatively insensitive to the starting errors. In fact, we need only that

$$|F_i - f(t_i)| = O(h^{p-1}), \qquad i = 0, 1, \ldots, r-1 \qquad (7.20)$$

to achieve an order of convergence p for a method whose order of consistency is p.

To use (7.19) to establish usable bounds on the error in the approximate solution is usually not very profitable. To compute the right-hand side of (7.19) we need to bound the consistency error. This requires some knowledge of the properties of the unknown solution $f(t)$ and it is consequently rather difficult to carry through. Furthermore, even if $|\varepsilon_n|$ can be so bounded, the results may be quite pessimistic, that is, several orders of magnitude larger than the actual error. The main use of (7.19) is in the qualitative information it carries. It establishes the convergence of a class of approximation methods and shows the dependence of the discretization error on the consistency error and the starting errors. This gives us some initial insight into the expected effectiveness of the methods.

Exercise 7.6. Show that if f and K are sufficiently smooth (that is, sufficiently differentiable), then the approximation method (7.3) is convergent with order 2.

Exercise 7.7. Show that both Simpson's method 1 and Simpson's method 2 are consistent of order 4 for sufficiently smooth f and K. Consequently, in the absence of starting errors, both methods have order of convergence 4.

7.4. Error estimates and numerical stability. While the order of convergence of a method is useful in establishing its potential efficiency, one frequently wants to have a better idea of how the error can be expected to behave. For example, it is of some practical concern to know whether the errors will vary smoothly from point to point or whether they oscillate more or less unpredictably within the bounds given by (7.19). To answer this question we need to obtain an *error estimate*, not just a bound. A closely related problem is the question of the growth of the error. If we consider (7.19) for fixed h, we see that the bound is essentially proportional to e^{WLt_n}. If WLt_n is much larger than one, then (7.19) allows for a rather large error even when h is quite small. Of course, since (7.19) is only a bound, the actual error may be much smaller, but one needs to consider whether there are some cases when the bound is realistic. If the actual solution grows like e^{WLt}, then there is no problem since the relative error will remain small. If, however, the solution grows more slowly than the error, then the method may have poor accuracy even though it is of high order. This phenomenon, which we call *numerical instability*, is well known in the solution of ordinary differential equations. As we shall see, this type of instability also occurs with Volterra equations.

In numerical analysis the term "stability" is used with several different meanings. Some authors make stability essentially synonymous with convergence. With an appropriate definition of the term one can show that, rather generally, consistency and stability imply convergence. This is not the definition we want to use here, since all reasonable methods for Volterra integral equations of the second kind are convergent if they are consistent. What we are concerned with here are cases where, for fixed h, the error can grow rapidly with t, while the solution itself does not grow very fast at all. We therefore use the term numerical instability for such a phenomenon. Actually, the above description is quite imprecise and the phrase "the error can grow rapidly" can be defined rigorously in several ways. Corresponding to these different definitions one obtains alternative ways of analyzing numerical stability. The method used in this section is the most general, but requires some detailed manipulations. A somewhat less technical approach will be described in § 7.5.

To simplify the arguments we consider only the linear case $K(t, s, f(s)) = k(t, s)f(s)$. The nonlinear case can be treated in much the same manner, yielding similar conclusions at the expense of additional work.

With the assumed linearity, the equation defining the error becomes

$$\varepsilon_n = h \sum_{i=0}^{n} w_{ni} k(t_n, t_i) \varepsilon_i - \delta(h, t_n), \qquad n = r, r+1, \ldots. \qquad (7.21)$$

The starting errors $\varepsilon_0 = \eta_0$, $\varepsilon_1 = \eta_1, \ldots, \varepsilon_{r-1} = \eta_{r-1}$ will be determined by the starting method used. The overall error, defined by (7.21), arises from two sources: one contribution coming from the propagation of the starting errors and a second one due to the consistency error. These two components behave quite differently, so that for further discussion it is convenient to separate them. Let ε_n^C and ε_n^S denote the respective solutions of

$$\varepsilon_n^C = h \sum_{i=0}^{n} w_{ni} k(t_n, t_i) \varepsilon_i^C - \delta(h, t_n), \qquad n = r, r+1, \ldots,$$
$$\varepsilon_0^C = \varepsilon_1^C = \cdots = \varepsilon_{r-1}^C = 0, \qquad (7.22)$$

and

$$\varepsilon_n^S = h \sum_{i=0}^{n} w_{ni} k(t_n, t_i) \varepsilon_i^S, \qquad n = r, r+1, \ldots,$$
$$\varepsilon_n^S = \eta_n, \qquad n = 0, 1, \ldots, r-1. \qquad (7.23)$$

We call ε_n^C the *accumulated consistency error* and ε_n^S the *accumulated starting error*. Their sum is the discretization error

$$\varepsilon_n = \varepsilon_n^C + \varepsilon_n^S, \qquad (7.24)$$

and is a solution of (7.21).

We now define more precisely what is meant by saying that the error varies smoothly.

DEFINITION 7.5. *For each admissible h, let $\xi_0(h), \xi_1(h), \ldots, \xi_N(h)$ be a sequence of numbers. We say that this set of sequences has an* expansion *of order p if there exists a continuous function $\chi(t)$ such that, for $n = 0, 1, \ldots, N$,*

$$\xi_n(h) = h^p \chi(t_n) + O(h^q), \tag{7.25}$$

where $q > p$.

If the discretization error has an expansion, then it varies smoothly from point to point in the sense that the scaled quantity ε_n/h^p tends to a limit which is a continuous function. The existence of such an expansion is closely related to a similar expansion for the numerical integration formula used.

Using Definition 7.5, we say that the local consistency error has an expansion of order p (for a given problem) if there exists a continuous function $Q(t)$ such that

$$\delta(h, t_n) = h^p Q(t_n) + O(h^q), \qquad q > p. \tag{7.26}$$

THEOREM 7.3. *Consider the linear equation*

$$f(t) = g(t) + \int_0^t k(t, s) f(s)\, ds.$$

If the local consistency error has an expansion of the form (7.26), *then the accumulated consistency error also has an expansion. In particular,*

$$\varepsilon_n^C = h^p e(t_n) + O(h^P), \tag{7.27}$$

where $P = \min(q, p+1)$ and $e(t)$ is the solution of

$$e(t) = -Q(t) + \int_0^t k(t, s) e(s)\, ds, \tag{7.28}$$

provided that all functions involved in (7.28) *are sufficiently smooth so that the method in question is consistent for* (7.28) *with order at least one.*

Proof. From (7.22) and (7.26)

$$\varepsilon_n^C = h \sum_{i=0}^n w_{ni} k(t_n, t_i) \varepsilon_i^C - h^p Q(t_n) + O(h^q), \qquad n = r, r+1, \ldots. \tag{7.29}$$

Now, by Definition 7.3,

$$h^p e(t_n) = -h^p Q(t_n) + h \sum_{i=0}^n w_{ni} k(t_n, t_i) h^p e(t_i) + h^p \delta_e(h, t_n),$$

$$n = r, r+1, \ldots, \tag{7.30}$$

where $\delta_e(h, t)$ is the local consistency error for (7.28). Subtracting (7.30) from (7.29), and using the fact that $\varepsilon_0^C = \varepsilon_1^C = \cdots = \varepsilon_{r-1}^C = 0$, we get

$$\varepsilon_n^C - h^p e(t_n) = h \sum_{i=0}^{n} w_{ni} k(t_n, t_i) \{\varepsilon_i^C - h^p e(t_i)\}$$

$$- h^{p+1} \sum_{i=0}^{r-1} w_{ni} k(t_n, y_i) e(t_i) - h^p \delta_e(h, t_n)$$

$$+ O(h^q), \qquad n = r, r+1, \ldots. \qquad (7.31)$$

Since by assumption $\delta_e(h, t)$ is at least $O(h)$, it follows from Theorem 7.1 that
$$\varepsilon_n^C - h^p e(t_n) = O(h^p),$$
which was to be proved. □

Exercise 7.8. Show that, for sufficiently smooth $f(t)$ and $k(t, s)$, the trapezoidal method described in § 7.1 has a local consistency error which has an expansion with $p = 2$ and $q = 4$.

It is also a straightforward exercise to show that for the methods suggested in § 7.2 (and indeed all standard numerical integration rules) the local consistency error has an expansion. Thus, in general, the accumulated consistency error behaves smoothly. The accumulated starting error, however, does not always behave as nicely and ε_n^S may exhibit oscillations of magnitude $O(\varepsilon_n^S)$. Whether or not this occurs seems to depend on the pattern in the weights w_{ni}; more specifically it is governed by the following concept.

DEFINITION 7.6. A method of type (7.7) is said to have a *repetition factor* ρ if ρ is the smallest integer such that

$$w_{n+\rho, i} = w_{ni} \qquad (7.32)$$

for $i = 0, 1, 2, \ldots, n - q$, where q is an integer independent of n.

According to this definition the trapezoidal method, the fourth order Gregory method, and Simpson's method 2 have a repetition factor of 1, while Simpson's method 1 has repetition factor 2.

THEOREM 7.4. *Consider the numerical method* (7.7) *under the assumptions*
 (i) *the method has repetition factor* 1,
 (ii) *the method is convergent with at least order* 1 *for equations* (7.35) *below,*
 (iii) *the starting errors* $\eta_i = F_i - f(t_i)$ *can be written as*

$$\eta_i = \gamma_i h^p + O(h^{p+1}), \qquad i = 0, 1, \ldots, r-1. \qquad (7.33)$$

Then the accumulated starting error ε_n^S *has an asymptotic expansion of the form*

$$\varepsilon_n^S = h^{p+1} \sum_{i=0}^{r-1} \gamma_i C_j(t_n) + O(h^{p+2}), \qquad (7.34)$$

where $C_j(t)$ is the solution of

$$C_j(t) = V_j k(t, t_j) + \int_0^t k(t,s) C_j(s)\, ds, \qquad j = 0, 1, \ldots, r-1. \tag{7.35}$$

The V_j's are constants whose value will be exhibited in the proof.

Proof. From (7.23) we see that ε_n^S is just a linear combination of the η_i. Let us write

$$\varepsilon_n^S = \sum_{j=0}^{r-1} c_{nj} \eta_j, \qquad n = r, r+1, \ldots, \tag{7.36}$$

then substitute this into (7.23), exchange orders of summation, and equate coefficients of η_j. This gives

$$c_{nj} = h w_{nj} k(t_n, t_j) + h \sum_{i=r}^n w_{ni} k(t_n, t_i) c_{ij}, \qquad j = 0, 1, \ldots, r-1.$$

Since the method has repetition factor 1 the weights w_{nj}, $j = 0, 1, \ldots, r-1$ are independent of n, provided $n \geq r + q$. We can therefore write $w_{nj} = V_j$ and get

$$c_{nj} = h V_j k(t_n, t_j) + h \sum_{i=r}^n w_{ni} k(t_n, t_i) c_{ij}. \tag{7.37}$$

If we now apply the approximation method to (7.35) and compare the resulting equation with (7.37), we see that

$$c_{nj} = h C_j(t_n) + O(h^2)$$

and (7.34) follows. □

We see then that for methods having a repetition factor of 1 both the accumulated consistency error and the accumulated starting error have an asymptotic expansion and therefore the total error behaves in a systematic way. For methods with repetition factor larger than one this is not necessarily so and significant oscillations may appear.

Next, let us consider the question of numerical stability, starting with a precise definition of the term. To provide some motivation, consider what happens when the right-hand side of (7.1) is perturbed by an amount $\Delta g(t)$. This will change the solution of the equation to $\hat{f}(t)$, given by

$$\hat{f}(t) = g(t) + \Delta g(t) + \int_0^t k(t,s) \hat{f}(s)\, ds. \tag{7.38}$$

The difference between f and $\hat{f}$ then satisfies

$$f(t) - \hat{f}(t) = -\Delta g(t) + \int_0^t k(t,s)\{f(s) - \hat{f}(s)\}\, ds. \tag{7.39}$$

Thus, the change in the solution satisfies an integral equation of the same form as the original equation, except for a difference in the nonhomogeneous term. Since there are invariably some inaccuracies present, we must always expect an error whose growth is governed by (7.39). If all errors are governed by such an equation, we call the corresponding method numerically stable. If the error can have a component which grows at a faster rate, we have a case of numerical instability. More precisely, we can make the following definition.

DEFINITION 7.7. A method is said to be *numerically stable* with respect to an error component ε_n if that error has an expansion of the form

$$\varepsilon_n = h^p e(t_n) + O(h^q), \qquad q > p > 0, \tag{7.40}$$

with $e(t)$ satisfying an equation

$$e(t) = Q(t) + \int_0^t k(t, s) e(s)\, ds, \tag{7.41}$$

for some $Q(t)$.

According to this definition and our previous results the trapezoidal method, the fourth order Gregory method and Simpson's method 2 are stable with respect to the accumulated consistency error as well as the accumulated starting error. Simpson's method 1 is stable with respect to ε_n^C, but not necessarily with respect to the accumulated starting error.

Theorem 7.3 implies that all standard methods are stable with respect to the accumulated consistency error. Theorem 7.4 shows the same thing for the accumulated starting errors, provided the method has repetition factor 1. The remaining question is the stability with respect to the accumulated starting error of methods having a repetition factor larger than one. The theoretical situation here is somewhat complicated, but the general result is that such methods have a tendency to show unstable error growth. Rather than trying to provide a general analysis, we will consider a specific case by looking at Simpson's method 1 in some detail.

Writing out (7.23) for Simpson's method 1 and rearranging, we have

$$\varepsilon_{2m}^S = u_{2m} + \tfrac{1}{3} hk(t_{2m}, t_2) \varepsilon_2^S$$

$$+ \tfrac{1}{3} h \sum_{i=2}^{m} \{k(t_{2m}, t_{2i-2}) \varepsilon_{2i-2}^S + 4k(t_{2m}, t_{2i-1}) \varepsilon_{2i-1}^S + k(t_{2m}, t_{2i}) \varepsilon_{2i}^S\}, \tag{7.42}$$

$$\varepsilon_{2m+1}^S = v_{2m+1} + \tfrac{9}{8} hk(t_{2m+1}, t_2) \varepsilon_2^S + \tfrac{3}{8} hk(t_{2m+1}, t_3) \varepsilon_3^S$$

$$+ \tfrac{1}{3} h \sum_{i=2}^{m} \{k(t_{2m+1}, t_{2i-1}) \varepsilon_{2i-1}^S + 4k(t_{2m+1}, t_{2i}) \varepsilon_{2i}^S$$

$$+ k(t_{2m+1}, t_{2i+1}) \varepsilon_{2i+1}^S\} \tag{7.43}$$

where

$$u_{2m} = \tfrac{1}{3}h\{k(t_{2m}, t_0)\eta_0 + 4k(t_{2m}, t_1)\eta_1\}, \qquad (7.44)$$

$$v_{2m+1} = \tfrac{3}{8}h\{k(t_{2m+1}, t_0)\eta_0 + 3k(t_{2m+1}, t_1)\eta_1\}. \qquad (7.45)$$

Since each set of summed terms in (7.42) and (7.43) represents an approximation to an integral, it appears that the expansion for ε_n^S, if it exists at all, will be given by a set of coupled integral equations. It is not hard to see that the appropriate system is

$$x(t) = u(t) + \tfrac{1}{3}\int_0^t k(t, s)x(s)\,ds + \tfrac{2}{3}\int_0^t k(t, s)y(s)\,ds, \qquad (7.46)$$

$$y(t) = v(t) + \tfrac{2}{3}\int_0^t k(t, s)x(s)\,ds + \tfrac{1}{3}\int_0^t k(t, s)y(s)\,ds, \qquad (7.47)$$

where

$$u(t) = \tfrac{1}{3}\{k(t, t_0)\eta_0 + 4k(t, t_1)\eta_1\}, \qquad (7.48)$$

$$v(t) = \tfrac{3}{8}\{k(t, t_0)\eta_0 + 3k(t, t_1)\eta_1\}. \qquad (7.49)$$

Once we have guessed this form, we can then proceed rigorously.

THEOREM 7.5. *If $x(t)$ and $y(t)$ satisfy (7.46)–(7.49), then*

$$\varepsilon_{2m}^S = hx(t_{2m}) + O(h^2), \qquad (7.50)$$

$$\varepsilon_{2m+1}^S = hy(t_{2m+1}) + O(h^2). \qquad (7.51)$$

Proof. We outline the steps only. Let X_n and Y_n denote the approximate solutions of (7.46) and (7.47) obtained by using the rectangular integration rules with stepsize $2h$. This yields

$$X_{2m} = u(t_{2m}) + \frac{2h}{3}\sum_{i=0}^{m-1} k(t_{2m, 2i})X_{2i} + \frac{4h}{3}\sum_{i=0}^{m-1} k(t_{2m, 2i})Y_{2i} \qquad (7.52)$$

for (7.46), and a similar equation for (7.47). By scaling and rearranging (7.42) we see that it differs from (7.52) only by terms of order h. A similar result holds for (7.47) and its approximation. But because of the integration rule chosen,

$$X_{2m} - x(t_{2m}) = O(h),$$

and (7.50) and (7.51) follow. □

Exercise 7.9. Carry out the details of the proof of Theorem 7.5.

We can see a little more clearly what (7.46) and (7.47) imply if we introduce

$$z_1(t) = x(t) + y(t), \qquad z_2(t) = x(t) - y(t),$$

and rearrange the equations. We then get

$$z_1(t) = u(t) + v(t) + \int_0^t k(t, s) z_1(s)\, ds, \tag{7.53}$$

$$z_2(t) = u(t) - v(t) - \tfrac{1}{3} \int_0^s k(t, s) z_2(s)\, ds. \tag{7.54}$$

The first component z_1 satisfies an equation of the form (7.41), but z_2 does not. Therefore, the method is not necessarily stable. To show that it can be unstable, we need to construct a case for which z_2 can grow much faster than the actual solution. Take the simple case

$$k(t, s) = -\lambda.$$

Using the results of § 1.2, we see that

$$z_1(t) = A_1(t) + B_1(t) e^{-\lambda t}, \tag{7.55}$$

$$z_2(t) = A_2(t) + B_2(t) e^{\lambda t/3} \tag{7.56}$$

where A_1, A_2, B_1, B_2 are functions whose exact form is immaterial for the argument. When $\lambda > 0$, the component z_2 of the accumulated starting error can grow exponentially. How the solution of the equation behaves depends on $g(t)$. For $g(t) = 1$, the solution is

$$f(t) = e^{-\lambda t},$$

and therefore decreases exponentially for $\lambda > 0$. Consequently, for large t the accumulated starting error may become so large as to obscure the nature of the solution altogether. The conclusion is then that Simpson's method 1 can show unstable error growth. In practice this method should be avoided.

It is possible to make a more complete analysis for methods with repetition factor higher than one, but this provides little further insight. All evidence is that such methods are not numerically stable and should be avoided.

7.5. Another view of stability. The discussion in the previous section focused on the dominant error term; a method is called numerically stable if the dominant error term satisfies an integral equation having the same kernel as the original equation. Certainly, if h is very small this dominant term closely represents the actual error, but it is not easy to say exactly what is meant by "very small". What one would really like to know is how the error behaves for practical values of h, say $h = 0.1$ or 0.05. This problem is too difficult to be solved in general so that some simplifying assumptions have to be made. One way to proceed is to take a very simple equation for which a complete analysis can be made. Therefore, let us consider the equation

$$f(t) = 1 - \lambda \int_0^t f(s)\, ds, \quad (7.57)$$

which has the solution $f(t) = e^{-\lambda t}$. Since for $\lambda > 0$ the solution decays exponentially, it is reasonable to require that the approximate solution should have the same behavior. We can make this more precise with the following definition.

DEFINITION 7.8. Let $F_n(h)$ denote the approximate solution of (7.57) computed by (7.7), with $\lambda > 0$, arbitrary starting values $F_0, F_1, \ldots, F_{r-1}$, and fixed h. Then the method (7.7) is said to be stable in the interval $(0, h_0)$ if for all $0 < h < h_0$

$$\lim_{n \to \infty} F_n(h) = 0. \quad (7.58)$$

The reason for defining stability with respect to the simple test equation (7.57) is that it now becomes possible to analyze fully the resulting discrete equations and thereby find the intervals of stability.

As a first example, let us take the trapezoidal method (7.3) with $g(t) = 1$ and $k(t, s) = -\lambda$. Taking the difference between (7.3) for n and (7.3) for $n - 1$ we get, dropping the argument h in $F_n(h)$,

$$F_n - F_{n-1} = -\tfrac{1}{2} h \lambda F_n - \tfrac{1}{2} h \lambda F_{n-1},$$

or

$$F_n = \frac{1 - \lambda h/2}{1 + \lambda h/2} F_{n-1}. \quad (7.59)$$

Therefore, if $\lambda > 0$ and $h > 0$, it follows that

$$\lim_{n \to \infty} F_n = 0$$

for all possible F_0. Therefore, the trapezoidal method is stable in $(0, \infty)$, an ideal situation which is often called *A-stability*.

Next, consider the fourth order Gregory method, with weights given by (7.8). Again, differencing the approximation formula gives

$$F_n - F_{n-1} = -\frac{3\lambda h}{8} F_n - \frac{19\lambda h}{24} F_{n-1} + \frac{5\lambda h}{24} F_{n-2} - \frac{\lambda h}{24} F_{n-3}.$$

This is a simple linear difference equation with constant coefficients and its solution can be written as

$$F_n = c_1 \rho_1^n + c_2 \rho_2^n + c_3 \rho_3^n, \quad (7.60)$$

where the c_i are constants determined by the starting values and ρ_1, ρ_2, ρ_3

are the roots of the polynomial

$$\left(1+\frac{3\lambda h}{8}\right)\rho^3 - \left(1-\frac{19\lambda h}{24}\right)\rho^2 - \frac{5\lambda h}{24}\rho + \frac{\lambda h}{24} = 0. \qquad (7.61)$$

Now, if $\max(|\rho_1|, |\rho_2|, |\rho_3|) < 1$, then we have again that

$$\lim_{n\to\infty} F_n = 0$$

for arbitrary starting values. Thus we see that the question of stability has been reduced to finding the roots of certain *characteristic polynomials*. If all of the roots are inside the unit circle in the complex plane, then the method is stable.

Let us look at (7.61) in a little more detail. For $h = 0$, the three roots are obviously 0, 0, 1. Since the roots of a polynomial are continuous functions of the coefficients, the roots of (7.61) for small h must be either near 0 or 1. We need be concerned only with roots near 1. When we write $\rho = (1+\varepsilon)$ and substitute in (7.61), we get

$$\left(1+\frac{3\lambda h}{8}\right)(1+\varepsilon)^3 - \left(1-\frac{19\lambda h}{24}\right)(1+\varepsilon)^2 - \frac{5\lambda h}{24}(1+\varepsilon) + \frac{\lambda h}{24} = 0.$$

Expanding and throwing away higher order terms (i.e., terms of order $2h\varepsilon$, ε^2, etc.) we find that

$$\varepsilon \simeq -\lambda h,$$

indicating that for small positive h all roots of (7.61) are inside the unit circle. Therefore the fourth order Gregory method is stable in some nonempty interval $(0, h_0)$.

Next, we ask for the largest interval in which the method is stable. This requires a detailed investigation, usually by numerical techniques, of the location of the roots as a function of h. When this is done, we find that the fourth order Gregory method is stable in the interval $(0, 3/\lambda)$.

A similar analysis can be made for any method of the form (7.7). In particular we find that Simpson's method 2 is stable in $(0, 2/\lambda)$. However, Simpson's method 1 is not stable for any $h > 0$.

Exercise 7.10. Investigate the stability regions for Simpson's method 1 and Simpson's method 2. Show that Simpson's method 2 is stable in $(0, 2/\lambda)$. Show that Simpson's method 1 is not stable for any $h > 0$.

The use of Definition 7.8 to define stability has many advantages. The approximate solution usually reduces to a linear difference equation with constant coefficients and hence its behavior is governed by the roots of the characteristic polynomials. It is then a simple computational matter to find the interval of stability. In practice, it is common to consider not only real values of λ, but to take λ in the complex plane with Re $(\lambda) > 0$. This allows for solutions with an oscillating component. Numerical root-finding methods can then be used to find the regions of stability in the complex plane. This has been done for many of the common methods.

It has, however, been argued that (7.57) is too simple and that the conclusions derived from it may be misleading. As a more representative example one might take

$$f(t) = 1 - \int_0^t \{\lambda + \mu(t-s)\}f(s)\,ds. \tag{7.62}$$

This equation satisfies the differential equation

$$f'' + \lambda f' + \mu f = 0 \tag{7.63}$$

and therefore has an exponentially decaying solution for $\lambda > 0$, $\mu > 0$. One can then say that a method is stable (for some value of h) if

$$\lim_{n \to \infty} F_n = 0,$$

where F_n is the solution obtained by using (7.7) on (7.62) with $\lambda > 0$ and $\mu > 0$.

The analysis of F_n is now more complicated, but still manageable. Again, as an example, consider the trapezoidal method (7.3). When applied to (7.62), we obtain after some manipulation

$$F_n - 2F_{n-1} + F_{n-2} = -\tfrac{1}{2}h\lambda F_n - h^2\mu F_{n-1} + \tfrac{1}{2}h\lambda F_{n-2}. \tag{7.64}$$

The solution of (7.64) is then

$$F_n = c_1 \rho_1^n + c_2 \rho_2^n,$$

where ρ_1 and ρ_2 are the roots of

$$(1+\tfrac{1}{2}h\lambda)\rho^2 - (2-h^2\mu)\rho + (1-\tfrac{1}{2}h\lambda) = 0,$$

that is

$$\rho_{1,2} = \frac{2 - h^2\mu \pm \sqrt{(2-h^2\mu)^2 - 4(1-h^2\lambda^2/4)}}{2(1+h\lambda/2)}. \tag{7.65}$$

It is no longer easy to see just how the roots behave as functions of λ and μ. However, for small λ we find that if $h^2\mu > 4$, then one of the roots of (7.65) will have magnitude greater than one, implying unstable error growth. Thus, if (7.62) is used as the test equation, the trapezoidal method is not stable for all values of h (that is, not A-stable).

If we compare the view of stability in this section with that of the previous section, we see that neither approach is completely satisfactory. The results of § 7.4 apply only as $h \to 0$ and hence give no information on the interval of stability. This difficulty is overcome by using something like Definition 7.7, but at the expense of restricting the analysis to a simple test case. As our discussion has shown, the form of this test case can affect the conclusions. This in turn raises the question to what extent (7.57) and (7.62) represent the general case. The answer to this question is not known at the present. Nevertheless, stability analysis does give some insight by characterizing some methods as stable and others as unstable. Unstable methods, such as Simpson's method 1, although convergent, may show some rapid error growth and should therefore be avoided. As long as we are looking only for this kind of qualitative information, it is quite immaterial which definition of stability we use.

7.6. Block-by-block methods. Except for some relatively low order methods, such as the trapezoidal method, the scheme (7.7) requires one or more starting values which must be found in some other way. The method which we now describe not only gives starting values but provides a convenient and efficient way for solving the equation over the whole interval.

The so-called *block-by-block* methods are a generalization of the well-known *implicit Runge–Kutta* methods for ordinary differential equations and one finds the latter term also used in connection with integral equations. The idea behind the block-by-block methods is quite general, but is most easily understood by considering a specific case.

Let us assume that we have decided to use Simpson's rule as the numerical integration formula. If we knew F_1, then we could simply compute F_2 by

$$F_2 = g(t_2) + \frac{h}{3}\{K(t_2, t_0, F_0) + 4K(t_2, t_1, F_1) + K(t_2, t_2, F_2)\}. \tag{7.66}$$

To obtain a value for F_1, we introduce another point $t_{1/2} = h/2$ and the corresponding value $F_{1/2}$, then use Simpson's rule with F_0, $F_{1/2}$, and F_1 to give

$$F_1 = g(t_1) + \frac{h}{6}\{K(t_1, t_0, F_0) + 4K(t_1, t_{1/2}, F_{1/2}) + K(t_1, t_1, F_1)\}. \tag{7.67}$$

To deal with the unknown value $F_{1/2}$ we approximate it by quadratic interpolation, using values F_0, F_1 and F_2, that is, we replace $F_{1/2}$ by

$$F_{1/2} \simeq \tfrac{3}{8}F_0 + \tfrac{3}{4}F_1 - \tfrac{1}{8}F_2,$$

so that we can compute F_1 by

$$F_1 = g(t_1) + \frac{h}{6}\{K(t_1, t_0, F_0) + 4K(t_1, t_{1/2}, \tfrac{3}{8}F_0 + \tfrac{3}{4}F_1 - \tfrac{1}{8}F_2) + K(t_1, t_1, F_1)\}. \tag{7.68}$$

Equations (7.66) and (7.68) are a pair of simultaneous equations for F_1 and F_2. For sufficiently small h there exists a unique solution which can be obtained by the method of successive substitution or by a more efficient procedure, such as Newton's method.

The general process should now be clear; for $m = 0, 1, 2, \ldots$ we compute the approximate solution by

$$F_{2m+1} = g(t_{2m+1}) + h \sum_{i=0}^{2m} w_i K(t_{2m+1}, t_i, F_i)$$

$$+ \frac{h}{6} \{K(t_{2m+1}, t_{2m}, F_{2m})$$

$$+ 4K(t_{2m+1}, t_{2m+1/2}, \tfrac{3}{8}F_{2m} + \tfrac{3}{4}F_{2m+1} - \tfrac{1}{8}F_{2m+2})$$

$$+ K(t_{2m+1}, t_{2m+1}, F_{2m+1})\}, \quad (7.69)$$

$$F_{2m+2} = g(t_{2m+2}) + h \sum_{i=0}^{2m} w_i K(t_{2m+2}, t_i, F_i)$$

$$+ \frac{h}{3} \{K(t_{2m+2}, t_{2m}, F_{2m})$$

$$+ 4K(t_{2m+2}, t_{2m+1}, F_{2m+1}) + K(t_{2m+2}, t_{2m+2}, F_{2m+2})\}, \quad (7.70)$$

where

$$\{w_i\} = \tfrac{1}{3}\{1, 4, 2, \ldots, 2, 4, 1\}, \quad i = 0, 1, \ldots, 2m,$$

and

$$t_{2m+1/2} = t_{2m} + \frac{h}{2}.$$

At each step (7.69) and (7.70) have to be solved simultaneously for the unknowns F_{2m+1} and F_{2m+2}, so that we obtain a block of unknowns at a time. This explains the origin of the term block-by-block method.

Let us briefly consider the question of convergence and stability for block-by-block methods. The analysis parallels the development in §§ 7.3 to 7.5, with some notational complications. Since there is not much new here we merely sketch the arguments. For simplification we consider only the linear case. Going through the usual procedure we find that the errors satisfy the equations

$$\varepsilon_{2m+1} = h \sum_{i=0}^{2m} w_i k(t_{2m+1}, t_i) \varepsilon_i$$

$$+ h\{c_{2m}\varepsilon_{2m} + c_{2m+2}\varepsilon_{2m+1} + c_{2m+2}\varepsilon_{2m+2}\} + R_{2m+1}, \quad (7.71)$$

$$\varepsilon_{2m+2} = h \sum_{i=0}^{2m} w_i k(t_{2m+2}, t_i) \varepsilon_i$$

$$+ h\{d_{2m}\varepsilon_{2m} + d_{2m+1}\varepsilon_{2m+1} + d_{2m+2}\varepsilon_{2m+2}\} + R_{2m+2}, \quad (7.72)$$

where the expression for c_i and d_i are obvious and

$$R_{2m+1} = h \sum_{i=0}^{2m} w_i k(t_{2m+1}, t_i) f(t_i)$$

$$+ h\{c_{2m}f(t_{2m}) + c_{2m+1}f(t_{2m+1}) + c_{2m+2}f(t_{2m+2})\}$$

$$- \int_0^{t_{2m+1}} k(t_{2m+1}, g) f(t)\, dt,$$

$$R_{2m+2} = h \sum_{i=0}^{2m} w_i k(t_{2m+2}, t_i) f(t_i)$$

$$+ h\{d_{2m}f(t_{2m}) + d_{2m+1}f(t_{2m+1}) + d_{2m+2}f(t_{2m+2})\}$$

$$- \int_0^{t_{2m+2}} k(t_{2m+2}, t) f(t)\, dt.$$

An elementary analysis then shows that for sufficiently smooth k and f the terms R_{2m+1} and R_{2m+2} are of order h^4, so that, applying Theorem 7.1 we find that

$$\max_{0 \leq i \leq N} |\varepsilon_i| = O(h^4), \tag{7.73}$$

showing that the method has order of convergence four.

Exercise 7.11. Find the expressions for c_i and d_i in (7.71) and (7.72).

Exercise 7.12. Show that for sufficiently smooth k and f

$$R_{2m+1} = \frac{h^5}{180} \sum_{j=0}^{m-1} \frac{d^4}{ds^4}\{k(t_{2m+1}, t_{2j})f(t_{2j})\}$$

$$- \frac{h^4}{96} k(t_{2m+1}, t_{2m+1/2}) f'''(t_{2m+1/2}) + O(h^5),$$

$$R_{2m+2} = \frac{h^5}{180} \sum_{j=0}^{m} \frac{d^4}{ds^4}\{k(t_{2m+2}, t_{2j})f(t_{2j})\} + O(h^5).$$

Use this to show that (7.73) holds.

To obtain error expansions we consider (7.71) and (7.72), proceeding as in §7.4. Taking into consideration the results of Exercise 7.12, we are led to consider the system

$$x(t) = R_1(t) + \tfrac{1}{3}\int_0^t k(t,s)x(s)\,ds + \tfrac{2}{3}\int_0^t k(t,s)y(s)\,ds, \qquad (7.74)$$

$$y(t) = R_2(t) + \tfrac{1}{3}\int_0^t k(t,s)x(s)\,ds + \tfrac{2}{3}\int_0^t k(t,s)y(s)\,ds, \qquad (7.75)$$

where

$$R_2(t) = \tfrac{1}{180}\int_0^t \frac{d^4}{ds^4}\{k(t,s)f(s)\}\,ds, \qquad (7.76)$$

$$R_1(t) = R_2(t) - \tfrac{1}{96}k(t,t)f'''(t). \qquad (7.77)$$

When (7.74) and (7.75) are discretized by the rectangular rule and compared with (7.71) and (7.72), it is found that

$$\varepsilon_{2m+1} = h^4 x(t_{2m+1}) + O(h^5), \qquad (7.78)$$

$$\varepsilon_{2m+2} = h^4 y(t_{2m+2}) + O(h^5). \qquad (7.79)$$

Exercise 7.13. Prove the results (7.78) and (7.79).

Since $x(t)$ and $y(t)$ are generally not the same, there does not exist an asymptotic expansion for the error at all points. But there is a separate expansion for the even and odd-numbered points.

In spite of the lack of a unique expansion for the error, the method can be considered stable with respect to the consistency error. (There are, of course, no starting errors.) To see this, subtract (7.74) from (7.75) to give

$$y(t) = x(t) + R_2(t) - R_1(t). \qquad (7.80)$$

When this is substituted into (7.74), the equation for $x(t)$ becomes

$$x(t) = R_1(t) + \tfrac{2}{3}\int_0^t k(t,s)\{R_2(s) - R_1(s)\}\,ds + \int_0^t k(t,s)x(s)\,ds. \qquad (7.81)$$

The equation for $x(t)$ then satisfies the stability criterion (7.41). A similar argument can be made for $y(t)$, so that the method can be considered stable according to Definition 7.7. It is also known that if Definition 7.8 is used as the stability criterion, then the method is stable for all values of h, that is, it is A-stable.

As shown by (7.78) and (7.79),

$$\varepsilon_{2m+2} - \varepsilon_{2m+1} = O(h^4), \qquad (7.82)$$

indicating that there exist oscillations of the order of magnitude as the error itself. This may be considered a minor drawback of the method, although it is easily eliminated by retaining, say, only the results at the even-numbered points.

The general procedure for constructing block-by-block methods should be easy to infer from what we have said, but becomes even more obvious if the derivation of (7.69) and (7.70) is motivated as follows. In each interval $[t_{2i}, t_{2i+2}]$, approximate $f(t)$ by a quadratic interpolating polynomial $P_2(t)$ on the points t_{2i}, t_{2i+1}, and t_{2i+2}. Replace $f(t)$ by this piecewise quadratic and integrate using Simpson's rule. The resulting numerical integration rule, when used to integrate between t_0 and t_{2m+1}, then between t_0 and t_{2m+2}, and expressed in terms of the values of $f(t_i)$, gives the approximating formulas (7.69) and (7.70). By using more points in each block, higher degree interpolating polynomials, and more accurate integration rules, one can construct block-by-block methods of arbitrarily high order.

7.7. Some numerical examples. The behavior predicted by the analysis in the previous sections is readily demonstrated with some simple examples. In addition to substantiating the theoretical predictions, such numerical experiments serve to give some additional insight into the relative effectiveness of the methods. In this section we summarize some numerical results on the equation considered in Example 7.1,

$$f(t) = e^t - \int_0^t e^{(t-s)} f(s) \, ds. \qquad (7.83)$$

While this equation is quite simple, the results are nevertheless typical of what one can expect in more complicated (but well-behaved) cases.

Example 7.2. Approximate solutions to (7.83) were computed using various fourth order methods. The observed errors are shown in Tables 7.2–7.5. In each case, any required starting values were taken as the exact solution $f(t) = 1$. Since the problem is linear, the unknowns at each step were determined by solving a linear equation.

These results indicate that all four methods have about the same accuracy, a situation which seems to hold true for many examples. Since all of the methods are fourth order, we expect a reduction of the error by a factor of $\frac{1}{16}$ as the stepsize is halved. This expectation is closely realized for the fourth order Gregory method, but less closely for the other three methods. An explanation for this observation can be found by looking at the column $h = 0.1$. (This is the only column where all computed results are listed.) For the fourth order Gregory method the error behaves quite smoothly in the sense that its graph is a smooth curve, but the errors for the other three methods do not behave quite as nicely. Certainly for the fourth order block-by-block method this is to be expected, since the errors on odd- and even-numbered points have a different expansion. For the two Simpson's methods the oscillations are due to the effect of the $O(h^5)$ terms which in this case are still significant. To understand why such higher order oscillations do not affect the fourth order Gregory method we must return to

TABLE 7.2
Errors in the solution of (7.83) by the fourth order Gregory method.

t	$h=0.1$	$h=0.05$	$h=0.025$
0.2	—	—	-1.7×10^{-9}
0.3	—	-4.0×10^{-8}	-2.8×10^{-9}
0.4	—	-5.7×10^{-8}	-3.8×10^{-9}
0.5	-1.1×10^{-6}	-7.4×10^{-8}	-4.9×10^{-9}
0.6	-1.4×10^{-6}	-9.1×10^{-8}	-5.9×10^{-9}
0.7	-1.7×10^{-6}	-1.1×10^{-7}	-7.0×10^{-9}
0.8	-2.0×10^{-6}	-1.3×10^{-7}	-8.0×10^{-9}
0.9	-2.3×10^{-6}	-1.4×10^{-7}	-9.1×10^{-9}
1.0	-2.5×10^{-6}	-1.6×10^{-7}	-1.0×10^{-8}

TABLE 7.3
Errors in the solution of (7.83) by Simpson's method 1.

t	$h=0.1$	$h=0.05$	$h=0.025$
0.1	—	-3.6×10^{-9}	-2.1×10^{-10}
0.2	-1.1×10^{-7}	-6.6×10^{-9}	-4.0×10^{-10}
0.3	-4.1×10^{-7}	-9.5×10^{-9}	-6.2×10^{-10}
0.4	-2.0×10^{-7}	-1.2×10^{-8}	-8.1×10^{-10}
0.5	-5.6×10^{-7}	-1.5×10^{-8}	-9.9×10^{-10}
0.6	-2.6×10^{-7}	-1.8×10^{-8}	-1.2×10^{-9}
0.7	-7.2×10^{-7}	-2.0×10^{-8}	-1.4×10^{-9}
0.8	-3.2×10^{-7}	-2.2×10^{-8}	-1.6×10^{-9}
0.9	-9.0×10^{-7}	-2.5×10^{-8}	-1.7×10^{-9}
1.0	-3.6×10^{-7}	-2.7×10^{-8}	-1.9×10^{-9}

TABLE 7.4
Errors in the solution of (7.83) by Simpson's method 2.

t	h = 0.1	h = 0.05	h = 0.025
0.1	—	-3.6×10^{-9}	-2.2×10^{-10}
0.2	-1.2×10^{-7}	-6.6×10^{-9}	-4.1×10^{-10}
0.3	-4.1×10^{-7}	-1.0×10^{-8}	-6.2×10^{-10}
0.4	-2.0×10^{-7}	-1.26×10^{-8}	-8.2×10^{-10}
0.5	-4.9×10^{-7}	-1.6×10^{-8}	-1.0×10^{-10}
0.6	-2.7×10^{-7}	-1.9×10^{-8}	-1.2×10^{-4}
0.7	-5.7×10^{-7}	-2.2×10^{-8}	-1.4×10^{-9}
0.8	-3.5×10^{-7}	-2.5×10^{-8}	-1.6×10^{-9}
0.9	-6.5×10^{-7}	-2.8×10^{-8}	-1.7×10^{-9}
1.0	-4.3×10^{-7}	-3.0×10^{-8}	2.1×10^{-9}

TABLE 7.5
Errors in the solution of (7.83) by the fourth order block-by-block method.

t	h = 0.1	h = 0.05	h = 0.025
0.1	-4.4×10^{-9}	-3.6×10^{-7}	-2.1×10^{-10}
0.2	-1.2×10^{-7}	-7.2×10^{-9}	$-4.3 \times 10^{\infty}$
0.3	-1.2×10^{-7}	-1.1×10^{-8}	-6.4×10^{-10}
0.4	-2.4×10^{-7}	-1.4×10^{-8}	-9.2×10^{-10}
0.5	-2.4×10^{-7}	1.8×10^{-8}	-1.1×10^{-9}
0.6	-3.5×10^{-7}	-2.2×10^{-8}	-1.3×10^{-9}
0.7	-3.6×10^{-7}	-2.5×10^{-8}	-1.5×10^{-9}
0.8	-4.7×10^{-7}	-2.9×10^{-8}	-1.8×10^{-9}
0.9	-4.8×10^{-7}	-3.2×10^{-8}	-1.9×10^{-9}
1.0	-5.9×10^{-7}	-3.6×10^{-8}	-2.2×10^{-9}

Theorem 7.3. Working along the lines indicated, it can be shown that if the local consistency error has an expansion with several terms of the form

$$\delta(h, t_n) = h^p Q_1(t_n) + h^q Q_2(t_n) + \cdots,$$

then (ignoring a few technical restrictions), the accumulated consistency error has an expansion of the form

$$\varepsilon_n^C = h^p e_1(t_n) + h^q e_2(t_n) + \cdots.$$

For the fourth order Gregory method (and also the trapezoidal rule), the local consistency error has an expansion consisting of several terms (depending on the smoothness of k and f); therefore, the higher order terms in ε_n^C also have a systematic behavior. For the two Simpson's methods, the use of the three-eighths rule at every second step limits the expansion to one term. Hence we can expect small oscillations due to the $O(h^5)$ term.

Example 7.3. The expected instability in Simpson's method 1 is not apparent from Table 7.3. However, it is present and becomes more pronounced as the solution is computed over a larger interval.

Equation (7.83) was integrated with $h = 0.2$ to $t = 7$, using the four methods as in Example 7.2. The errors are listed in Table 7.6.

As can be seen from Table 7.6, Simpson's method 1 shows some noticeable oscillations. The errors in the stable methods are significantly smaller.

Actually, the objection could be raised that, since exact starting values were used, Simpson's method 1 should show no instability. As we have shown, the method is unstable only with respect to the starting errors! Such an argument ignores a few fine points. First, the higher order error components are still noticeable for $h = 0.2$ and these can contribute to instability. Second, as in all computations, there are round-off errors. At each step, the round-off error acts like a starting error for the next step. Hence, the

TABLE 7.6
Errors in the solution of (7.83) for large t.

t	4th order Gregory	Simpson 1	Simpson 2	4th order block-by-block
6.0	-2.9×10^{-4}	6.2×10^{-3}	-1.5×10^{-5}	-6.0×10^{-5}
6.2	-3.0×10^{-4}	-8.1×10^{-3}	-2.6×10^{-5}	-6.1×10^{-5}
6.4	-3.1×10^{-4}	1.1×10^{-2}	-1.6×10^{-5}	-6.4×10^{-5}
6.6	-3.2×10^{-4}	-1.4×10^{-2}	-2.7×10^{-5}	-6.5×10^{-5}
6.8	-3.3×10^{-4}	1.8×10^{-2}	-1.7×10^{-5}	-6.8×10^{-5}
7.0	-3.4×10^{-4}	-2.3×10^{-2}	-2.8×10^{-5}	-6.9×10^{-5}

propagation of round-off error is governed by the equation for the starting error. Even though small it can, as in this example, grow and eventually become large.

7.8. Explicit Runge–Kutta methods. In the solution of ordinary differential equations, a class of algorithms called *explicit Runge–Kutta* methods is popular in practice and has been studied in great depth. To develop such methods, one first writes down a general form for computing the solution at $t_n + h$ in terms of the solution at previous points. The general form involves a number of undetermined parameters which are chosen by performing a Taylor expansion in h on the exact and approximate solutions, then matching as many terms as possible. Since in general, the set of equations is underdetermined, many different methods are possible. A similar idea can be used for integral equations; since there is even more freedom here, the class of methods so obtained is very large.

To present a fairly comprehensive approach, we introduce the intermediate values

$$F_{nq} = \varphi_n(t_n + \theta_q h) + h \sum_{s=0}^{q-1} A_{qs} K(t_n + d_{qs} h, t_n + c_s h, F_{ns}), \quad (7.84)$$

for $q = 1, 2, \ldots, p$. Using $\varphi_n(0) = F_{00} = g(0)$, this equation can be used to determine the sequence of values $F_{00}, F_{01}, \ldots, F_{0p}, F_{10}, F_{11}, \ldots, F_{1p}, \ldots$ The value F_{np} is taken as an approximation to $f(t_n + h)$; also $F_{np} = F_{n+1,0}$.

The function $\varphi_n(t)$ is chosen so that it is an approximation to

$$g(t) + \int_0^{t_n} K(t, s, f(s)) \, ds,$$

while the parameters θ_q, A_{qs}, d_{qs}, c_s are determined by a Taylor expansion. Actually, (7.84) is so general that only few restricted forms have been investigated in any detail. For example, by setting

$$\theta_q = d_{qs} = c_q, \quad (7.85)$$

we obtain a set of formulas sometimes called *Pouzet-type*. By taking

$$\theta_q = c_q, \quad (7.86)$$

$$d_{qs} = d_q, \quad (7.87)$$

we get the so-called *Beltyukov* methods.

For $p = 4$, a particular set of parameters for a Pouzet-type method is

$$\theta_0 = 0, \quad \theta_1 = \theta_2 = \tfrac{1}{2}, \quad \theta_3 = \theta_4 = 1,$$
$$A_{10} = A_{21} = \tfrac{1}{2},$$
$$A_{20} = A_{30} = A_{31} = 0,$$
$$A_{40} = A_{43} = \tfrac{1}{6},$$
$$A_{41} = A_{42} = \tfrac{1}{3},$$
$$A_{32} = 1.$$

The determination of the coefficients by Taylor expansion is a tedious matter. In many instances, such as for certain Pouzet-type formulas, there is an easier and more intuitive way of deriving the formulas. We can consider F_{nq} as an approximation to $f(t_n + qh)$. The first term on the right-hand side of (7.84) is then an approximation to

$$g(t_n + \theta_q h) + \int_0^{t_n} K(t_n + \theta_q h, s, f(s))\,ds, \tag{7.88}$$

while the second term represents

$$\int_{t_n}^{t_n+\theta_q h} K(t_n + \theta_q h, s, f(s))\,ds. \tag{7.89}$$

Once we have chosen a set of θ_i, with $0 \le \theta_i \le 1$, we can use standard quadrature rules to get the parameters A_{qs}. For example, take $p = 3$ and $\theta_1 = \frac{1}{3}$, $\theta_2 = \frac{2}{3}$, $\theta_3 = 1$, with the numerical integration formulas

$$\int_0^{h/3} f(s)\,ds \simeq \frac{h}{3} f(0), \tag{7.90}$$

$$\int_0^{2h/3} f(s)\,ds \simeq \frac{2h}{3} f\!\left(\frac{h}{3}\right), \tag{7.91}$$

$$\int_0^h f(s)\,ds = \frac{h}{4} f(0) + \frac{3h}{4} f\!\left(\frac{2h}{3}\right). \tag{7.92}$$

Using (7.90)–(7.92) to replace the integrals in (7.88) and (7.89), we are led to the formulas

$$F_{n1} = g\!\left(t_n + \frac{h}{3}\right) + \frac{h}{4} \sum_{i=0}^{n-1} \left\{ K\!\left(t_n + \frac{h}{3}, t_i, F_{i-1,3}\right) + 3K\!\left(t_n + \frac{h}{3}, t_i + \frac{2h}{3}, F_{i2}\right) \right\}$$

$$+ \frac{h}{3} K\!\left(t_n + \frac{h}{3}, t_n, F_{n-1,3}\right), \tag{7.93}$$

$$F_{n2} = g\!\left(t_n + \frac{2h}{3}\right) + \frac{h}{4} \sum_{i=0}^{n-1} \left\{ K\!\left(t_n + \frac{2h}{3}, t_i, F_{i-1,3}\right) + 3K\!\left(t_n + \frac{2h}{3}, t_i + \frac{2h}{3}, F_{i2}\right) \right\}$$

$$+ \frac{2h}{3} K\!\left(t_n + \frac{2h}{3}, t_n + \frac{h}{3}, F_{n1}\right), \tag{7.94}$$

$$F_{n3} = g(t_n + h) + \frac{h}{4} \sum_{i=0}^{n} \left\{ K(t_n + h, t_i, F_{i-1,3}) + 3K\!\left(t_n + h, t_i + \frac{2h}{3}, F_{i2}\right) \right\}, \tag{7.95}$$

with $F_{-1,3} = g(0)$.

In Table 7.7, equation (7.83) was solved using several values of h. The indicated order of convergence is three, an observation which is easily proved.

TABLE 7.7
Observed error in the solution of equation (7.83) by the explicit Runge–Kutta method (7.93)–(7.95).

t	$h = 0.2$	$h = 0.1$	$h = 0.05$
0.2	7.8×10^{-5}	9.0×10^{-6}	1.1×10^{-6}
0.4	1.5×10^{-4}	1.8×10^{-5}	2.2×10^{-6}
0.6	2.3×10^{-4}	2.7×10^{-5}	3.3×10^{-6}
0.8	3.1×10^{-4}	3.6×10^{-5}	4.3×10^{-6}
1.0	3.9×10^{-4}	4.5×10^{-5}	5.4×10^{-6}

Convergence and stability analyses for explicit Runge–Kutta methods are very similar to what has been presented in preceding sections of this chapter. The selection of the parameters assures local consistency and establishes its order. Theorem 7.2 can then be used to prove convergence. The repetition factor depends on the choice of the parameters, specifically on the numerical integration rule used to approximate (7.88). For (7.93)–(7.95), looking at $F_{n3} = F_{n+1}$ only, we see that it has a repetition factor of one and is therefore stable if the definition of § 7.4 is used. If the approach of § 7.5 is preferred, an application of the approximating formulas to the test equation (7.57) gives the characteristic polynomial from which the stability region can be computed. Thus, apart from technical considerations, the analysis of the explicit Runge–Kutta methods raises no new questions.

Whether explicit Runge–Kutta methods are widely useful in practice is an open question. On the whole, they appear to be less efficient than either the step-by-step methods of § 7.2 or the implicit Runge–Kutta methods, requiring more work for the same accuracy. The situation here is not quite analogous to the differential equations case. For differential equations the work required to solve nonlinear implicit formulas is often the main part of the work. Thus, explicit Runge–Kutta methods are reasonably competitive with implicit schemes, such as certain multistep methods. For integral equations, however, most of the work involves the repeated approximation of integrals; the additional work to solve the nonlinear equations is relatively unimportant. Therefore, implicit methods with their greater accuracy and better stability properties seem preferable. It is of course possible that explicit Runge–Kutta methods are the most suitable in special circumstances, but no definitive study on this has yet been carried out.

7.9. A summary of related ideas and methods. The basic framework established in this chapter allows for a great many variations. A considerable amount of work has been done on this topic and it is not our intention to review all of it here. Much of the work is, in any case, in the nature of variations on a general theme. In this section we survey some of the ideas which have received attention. Details can be found in the literature.

There is, as previously pointed out, a close connection between Volterra integral equations and initial value problems for ordinary differential equations. It is therefore not surprising that numerical methods for Volterra equations usually turn out to be generalizations of corresponding methods for differential equations. Consider, for example, the simple Volterra equation

$$f(t) = 1 + \int_0^t f(s)\, ds \qquad (7.96)$$

which is equivalent to the differential equation

$$f'(t) = f(t). \qquad (7.97)$$

Apply now the fourth order Gregory method to (7.96) and difference the resulting equation (7.7). Then

$$F_n - F_{n-1} = \frac{h}{24}\{9F_n + 21F_{n-1} - 5F_{n-2} + F_{n-3}\}. \qquad (7.98)$$

But this is just a standard linear multistep method for ordinary differential equations applied to (7.97).

In general, the approximation method (7.7), using either Gregory or Newton–Cotes integration, will reduce to a linear multistep method for ordinary differential equations. Because of this, many of the ideas developed in connection with differential equations have been extended to integral equations. One example is the definition of stability given in Definition 7.8. This is completely analogous to the stability definition usually used in differential equations. Another concept, which can be generalized is the so-called *predictor-corrector* method which, in differential equations, is used to overcome the implicit nature of certain multistep methods such as (7.98). To see how this can be done for integral equations write (7.1) as

$$f(t) = G_n(t) + Z_n(t), \qquad t_n \leq t \leq t_{n+1}, \qquad (7.99)$$

where

$$G_n(t) = g(t) + \int_0^{t_n} k(t, s, f(s))\, ds, \qquad (7.100)$$

$$Z_n(t) = \int_{t_n}^t k(t, s, f(s))\, ds. \qquad (7.101)$$

Then clearly

$$f(t_{n+1}) = G_n(t_{n+1}) + Z_n(t_{n+1}). \qquad (7.102)$$

If we know values $F_0, F_1, \ldots, F_n$, then $G_n(t_{n+1})$ can be approximated by a numerical integration using F_i instead of $f(t_i)$. To approximate $Z_n(t_{n+1})$ we can use quadratures which involve values outside the interval $[t_n, t_{n+1}]$. For example, we can use the formula

$$\int_{t_n}^{t_{n+1}} \varphi(t) \approx \frac{h}{24} \{55\varphi(t_n) - 59\varphi(t_{n-1}) + 37\varphi(t_{n-2}) - 9\varphi(t_{n-3})\}. \quad (7.103)$$

This integration rule can be derived by approximating $\varphi(t)$ by a cubic interpolating polynomial on the points $t_n, t_{n-1}, t_{n-2}, t_{n-3}$, then integrating.

If we apply (7.103) to (7.101), we are led to an approximation to F_{n+1} of the form

$$F_{n+1}^* = \hat{G}_{n+1} + \frac{h}{24} \{55K(t_{n+1}, t_n, F_n) - 59K(t_{n+1}, t_{n-1}, F_{n-1})$$

$$+ 37K(t_{n+1}, t_{n-2}, F_{n-2}) - 9K(t_{n+1}, t_{n-3}, F_{n-3})\} \quad (7.104)$$

where $\hat{G}_{n+1}$ is an approximation to $G_n(t_{n+1})$, generally of the form

$$\hat{G}_{n+1} = g(t_{n+1}) + h \sum_{j=0}^{n} w_{nj} K(t_{n+1}, t_j, F_j). \quad (7.105)$$

Equation (7.104) defines F_{n+1}^* explicitly in terms of $F_0, F_1, \ldots, F_n$ and is a predictor for F_{n+1}. To use this, we employ another integration rule, now involving also $\varphi(t_{n+1})$. For example, one might use

$$\int_{t_n}^{t_{n+1}} \varphi(t) \, dt \approx \frac{h}{24} \{9\varphi(t_{n+1}) + 19\varphi(t_n) - 5\varphi(t_{n-1}) + \varphi(t_{n-2})\}. \quad (7.106)$$

We take this to approximate $Z_n(t_{n+1})$, using the predicted value F_{n+1}^* in place of $f(t_{n+1})$. The corrected and final value for F_{n+1} is then computed by

$$F_{n+1} = \hat{G}_{n+1} + \frac{h}{24} \{9K(t_{n+1}, t_{n+1}, F_{n+1}^*) + 19K(t_{n+1}, t_n, F_n)$$

$$- 5K(t_{n+1}, t_{n-1}, F_{n-1}) + K(t_{n+1}, t_{n-2}, F_{n-2})\}. \quad (7.107)$$

Equations (7.104) and (7.107) are then a predictor-corrector scheme for the solution of (7.1). The scheme is very similar to the corresponding formulas for differential equations.

Another approach to solving Volterra equations of the second kind is based on an observation made in Chapter 1, that for degenerate kernels the integral equation can be reduced to a system of differential equations. When the kernel is simple enough to be closely approximated by a degenerate kernel with a few functions, this can be very efficient. How competitive in general such a method is with the more usual approaches is an unresolved question.

Finally, there are a variety of other methods, which, in one way or another, combine and modify the ideas outlined. All of these seem to have desirable features in some instances, but a comparative study is lacking.

Notes on Chapter 7. Texts on elementary numerical analysis rarely mention integral equations. An exception is an early book by Collatz [69], which discusses the use of standard numerical integration rules on integral equations. A more up-to-date discussion can be found in Churchhouse [67].

There are also few books specifically devoted to the numerical solution of integral equations. By far the most complete treatment is given by Baker in [17]. Beyond this, there are a number of collections of papers and conference proceedings dealing with numerical methods for integral equations [2], [11], [22], [84], [116], [205]. A complete bibliography of work up to 1971 is given in [197]; later surveys are by Tsalyuk [230] and Brunner [59].

Some early results on the approximate solution of Volterra equations can be found in [104], [181] and [196]. However, a concerted effort to develop a theory started only with the work of Pouzet [210], [211], [212]. Pouzet's primary interest lay in explicit Runge–Kutta methods, and his work resulted in a variety of algorithms, such as the one given in § 7.8. The development of the general approach used here, and in most subsequent work on Volterra integral equations, began with the work of Kobayashi [152] and Linz [163].

The question of numerical stability was first raised in [152]. The concept of a repetition factor was introduced and exploited in [163]. A later analysis by Noble [198] clarified and extended the original idea. The analysis given in § 7.4 was adapted from [198]. Conditions required for the existence of error expansions with several terms were studied by Hock ([126], [127]). Investigations of stability as defined in § 7.5 are given in [18], [20], [132], [133], [178], [185], [243]. The stability question continues to be of interest. Recent advances include the introduction of an asymptotic repetition factor [245] and the use of more complicated test equations [174].

Block-by-block methods seem to have been suggested first by Young [249]. The idea was developed in [63], [107], [111], [163], and analyzed in [83] and [164]. Brunner, in a series of papers, explores the closely related idea of piecewise polynomial approximation and collocation ([42], [51], [57], [58]). Additional methods, based on piecewise polynomial and spline functions, are studied in [91], [92], [93], [96], [97], [136], [186], [195].

Special starting methods are investigated in a number of papers (for example, [62], [76], [77], [140], [157], [199]). Garey [108] studied predictor-corrector type methods. The fact that some formulas for Volterra equations reduce to linear multistep methods for ordinary differential equations is useful in the study of their stability. Details on this can be found in [244].

In addition to the work of Pouzet, there has been considerable interest in explicit Runge–Kutta methods and various modifications. Beltyukov [28] gives some early work. Later studies were carried out by Garey [112], Baker and his co-workers ([16], [19], [21]), van der Houwen ([131], [134]), and Brunner [60].

The use of degenerate kernel approximations in the numerical solution of Volterra equations is due primarily to Bownds and his co-workers ([31], [34], [35], [36], [117], [246]). For a related idea, see [27].

There are many other papers which discuss various minor modifications of the ideas contained in this chapter. Occasionally, slightly different forms of the equation are used. For example, see [61], [98], [99], [141], [150], [151], [156], [175], [202], [203], [209], [215], [223], [224], [238], [241], [242]. Somewhat more unusual suggestions, using computers for algebraic manipulations, were made by Goldfine [118] and Stoutemeyer [225].

Chapter 8

Product Integration Methods for Equations of the Second Kind

The methods in Chapter 7 were developed under the assumption that all functions involved in the equations were well-behaved; in particular, we assumed that $K(t, s, f)$ and $g(t)$ were continuous. In some cases, the analysis also required additional smoothness for these functions. In many practical situations such conditions are not satisfied. If the kernel or some of its lower derivatives are unbounded, new methods are needed. As we shall see, the methods previously described can be adapted to these more complicated conditions if we use more powerful numerical integration techniques. One such technique is product integration which is briefly described in § 8.1.

Using the product integration method the extension of the standard techniques is conceptually quite simple. Unfortunately technical complications arise which make the analysis somewhat tedious. As a result a complete generalization of the results in the previous chapter has not been carried out, although it is fairly clear that, given sufficient patience, this can be done. In the interest of clarity and conciseness we will make our treatment somewhat more descriptive and less rigorous than that in Chapter 7. However, we present a complete discussion where this is possible without undue complication.

The equation considered here is

$$f(t) = g(t) + \int_0^t p(t, s) K(t, s, f(s))\, ds, \qquad 0 \le t \le T, \tag{8.1}$$

where we assume that

(i) $g(t)$ and $K(t, s, f)$ are continuous in the region of interest,

(ii) the well-behaved part of the kernel satisfies a Lipschitz condition of the form

$$|K(t, s, y) - K(t, s, z)| \le L\, |y - z|. \tag{8.2}$$

Since $K(t, s, f)$ is assumed to be smooth, all of the singular properties of the full kernel must be included in the term $p(t, s)$. However, some restrictions must be put on $p(t, s)$. We will use here the conditions of Theorem 4.8, so that a unique continuous solution exists.

After outlining the product integration technique in § 8.1, we begin with a rather simple numerical method which we call Euler's method in analogy with the corresponding method in ordinary differential equations. Because of its simplicity, this method has rather low accuracy, but the situation can be improved by the use of certain extrapolation techniques. In § 8.3 we introduce a more accurate method based on the product integration analogue to the trapezoidal method. Here some practical difficulties appear which are absent in the case of continuous kernels. In § 8.4 we show, with a specific example, how one can generalize the previously introduced block-by-block methods. In the concluding section of this chapter a theoretical justification for all of these methods is presented by establishing a general convergence theorem.

8.1. Product integration. The standard numerical integration rules, such as the trapezoidal and Simpson's methods, are constructed under the assumption that the integrand is at least bounded. When this is not the case such methods may not work. Even if the integrand is continuous, a great deal of accuracy is lost if higher derivatives fail to exist. Special methods are needed to handle such cases efficiently.

One of the most powerful ways to deal with poorly behaved integrands is *product integration*. To understand the motivation behind product integration let us briefly review the way in which the usual interpolatory integration rules are constructed. To evaluate

$$I = \int_a^b \varphi(t)\, dt \tag{8.3}$$

numerically, we first replace φ by some approximation $\hat{\varphi}$, then compute

$$\hat{I} = \int_a^b \hat{\varphi}(t)\, dt. \tag{8.4}$$

The approximation $\hat{\varphi}$ has to be chosen in such a way that the integral for $\hat{I}$ can be evaluated explicitly; usually $\hat{\varphi}$ will be some piecewise polynomial approximation to φ. For example, if $\hat{\varphi}$ is a piecewise linear approximation to φ obtained by interpolation on the points $a = t_0 < t_1 < \cdots < t_n = b$, $t_i - t_{i-1} = h$, then (8.4) is the (composite) trapezoidal rule. If we take as $\hat{\varphi}$ a piecewise quadratic interpolation on equidistant points, we obtain Simpson's rule.

A similar approach is taken to construct product integration rules but instead of approximating the whole integrand by a piecewise polynomial, we only approximate the well-behaved part. We write the integral as

$$I = \int_a^b p(t)\psi(t)\,dt, \tag{8.5}$$

where $\psi(t)$ is assumed to be continuous and generally smooth. Thus, whatever singularities or poor behavior the integrand has are included in $p(t)$. We then approximate ψ by a function $\hat{\psi}$ and compute

$$\hat{I} = \int_a^b p(t)\hat{\psi}(t)\,dt. \tag{8.6}$$

Again, the type of approximation must be chosen so that the integral in (8.6) can be evaluated. Typically, if we use piecewise polynomial approximations, then we must be able to evaluate (either explicitly or by an efficient numerical technique) integrals of the form

$$\int t^i p(t)\,dt. \tag{8.7}$$

For most problems of practical interest this can be done. From (8.5) and (8.6) we immediately get that

$$|I - \hat{I}| \leq \max_{a \leq t \leq b} |\psi(t) - \hat{\psi}(t)| \int_a^b |p(t)|\,dt, \tag{8.8}$$

so that error bounds and orders of convergence for product integration follow from standard results of approximation theory. For example, for a piecewise linear approximation to a smooth function ψ, $|\psi(t) - \hat{\psi}(t)| = O(h^2)$, so that the product trapezoidal is second order. By the same reasoning the product Simpson's rule, which is based on a piecewise quadratic approximation, is at least third order. We stress the words "at least", since occasionally one does achieve a higher order. This is well known for the standard Simpson's rule, where an $O(h^3)$ approximation of the integrand yields an integration method whose order is four. For the product Simpson's rule the situation is not as favorable and the order of convergence is not necessarily four. A detailed analysis for some typical values of $p(t)$, such as $p(t) = t^{-1/2}$, shows an order of convergence between three and four.

The explicit form of product integration formula depends of course on the method of approximation we use to construct $\hat{\psi}(t)$. Once a form has been chosen the rest is elementary, although occasionally tedious, algebra. In subsequent sections of this chapter the reader will find several cases where the details have been worked out; here we only describe a general scheme which includes both the product trapezoidal and product Simpson's rule. We subdivide the interval $[a, b]$ into n equal subintervals of width h, using $a = t_0 < t_1 < \cdots < t_n = b$, $t_i = a + ih$. In each subinterval we introduce a further subdivision, with points $t_i \leq t_{i0} \leq t_{i1} < \cdots < t_{im} \leq t_{i+1}$. The function ψ

will be approximated by a polynomial of degree m in each subinterval $[t_i, t_{i+1}]$, using a Lagrange-type interpolation on the points $t_{i0}, t_{i1}, \ldots, t_{im}$. From elementary interpolation theory we know that this piecewise polynomial can be explicitly represented as

$$\hat{\psi}(t) = \sum_{j=0}^{m} l_{ij}(t)\psi(t_{ij}), \qquad t_{i0} \leq t \leq t_{im}, \tag{8.9}$$

where the l_{ij} are the fundamental polynomials defined as

$$l_{ij}(t) = \frac{(t-t_{i0})\cdots(t-t_{i,j-1})(t-t_{i,j+1})\cdots(t-t_{im})}{(t_{ij}-t_{i0})\cdots(t_{ij}-t_{i,j-1})(t_{ij}-t_{i,j+1})\cdots(t_{ij}-t_{im})}. \tag{8.10}$$

Inserting this into (8.6), we have immediately that

$$\hat{I} = \sum_{i=0}^{n-1} \sum_{j=0}^{m} w_{ij}\psi(t_{ij}), \tag{8.11}$$

where

$$w_{ij} = \int_{t_i}^{t_{i+1}} p(t) l_{ij}(t)\, dt. \tag{8.12}$$

If the integrals in (8.7) are known, then the weights w_{ij} can be found explicitly.

8.2. A simple method for a specific example. As a simple example of (8.1) we take

$$p(t, s) = (t-s)^{-1/2}, \tag{8.13}$$

a case which is of some practical importance. To apply product integration we approximate the integrand by a piecewise constant function agreeing with the integrand at the left endpoint of each subinterval. This leads of course to the product integration analogue to the rather crude rectangular method in ordinary numerical integration. In the context of the integral equation we get the approximation

$$\int_0^{t_n} \frac{K(t_n, s, f(s))}{\sqrt{t_n - s}} ds \approx \sum_{j=0}^{n-1} K(t_n, t_j, f(t_j)) \int_{t_j}^{t_{j+1}} \frac{ds}{\sqrt{t_n - s}}. \tag{8.14}$$

Using this in (8.1), satisfying the resulting equation at $t_1, t_2, \ldots$, and denoting by F_n the approximation to $f(t_n)$, we obtain

$$F_n = g(t_n) + \sum_{j=0}^{n-1} w_{nj} K(t_n, t_j, F_j), \qquad n = 1, 2, \ldots, \tag{8.15}$$

where

$$w_{nj} = 2\{\sqrt{t_n - t_j} - \sqrt{t_n - t_{j+1}}\}. \tag{8.16}$$

Starting with $F_0 = g(0)$, successive values for F_n can be immediately computed from (8.15).

The product integration rule, being based on a piecewise constant approximation of the integrand, has, according to the discussion of the preceding section, an error of magnitude $O(h)$. If this order of magnitude is preserved in the solution of the integral equation (and we will see below that this is generally true), then we can expect that

$$|F_n - f(t_n)| = O(h). \qquad (8.17)$$

The method, which is usually called *Euler's method*, can therefore be expected to have low accuracy for reasonably sized h. Nevertheless, it is of some practical interest because of the simplicity of the weights w_{nj} and the ease with which the resulting system (8.15) can be solved. As we shall shortly see, higher order methods come with certain complications which are absent here. The situation is of course familiar to anyone with practical experience. Often the simpler methods are easy to use but yield poor results, while the more accurate methods involve some nontrivial technical complications. One way around this dilemma is to use certain extrapolation techniques in which quite accurate answers can be obtained using only results from the inaccurate methods.

One such technique is the so-called *Richardson's extrapolation*. The scope of this process is extensive, but we will discuss it only in the context of the problem at hand. Suppose we know not only that (8.17) is true, but that

$$F_n - f(t_n) = e(t_n)h + O(h^p), \qquad p > 1, \qquad (8.18)$$

where $e(t)$ is a function independent of h. By computing an approximate solution with two different stepsizes we can then eliminate the $O(h)$ term from the error and thus obtain a more accurate answer. To make this a little clearer, let $Y(t, h)$ denote the approximation to $f(t)$ using stepsize h. With t as one of the points of subdivision, we have

$$Y(t, h) = f(t) + e(t)h + O(h^p). \qquad (8.19)$$

Repeating the computation with stepsize $h/2$ gives

$$Y\left(t, \frac{h}{2}\right) = f(t) + \tfrac{1}{2}e(t)h + O(h^p). \qquad (8.20)$$

From these two equations we see that

$$f(t) = 2Y\left(t, \frac{h}{2}\right) - Y(t, h) + O(h^p). \qquad (8.21)$$

Thus, the "extrapolated" value $2Y(t, h/2) - Y(t, h)$ has an accuracy of order h^p, with $p > 1$.

In order to apply Richardson's extrapolation, one needs to know the exact order of the dominant error term. If this is not known, one can use a technique due to *Aitken* to get around the difficulty. Suppose we know only that

$$Y(t, h) = f(t) + e(t)h^q + O(h^p), \quad p > q, \quad (8.22)$$

with q unknown. Then, recomputing the solution with stepsize $h/2$ and $h/4$, we have

$$Y\left(t, \frac{h}{2}\right) = f(t) + e(t)\left(\frac{h}{2}\right)^q + O(h^p), \quad (8.23)$$

$$Y\left(t, \frac{h}{4}\right) = f(t) + e(t)\left(\frac{h}{4}\right)^q + O(h^p). \quad (8.24)$$

Ignoring the $O(h^p)$ terms and eliminating $f(t)$ and $e(t)$ from (8.22)–(8.24), we find that

$$q \simeq \log_2 \frac{Y(t, h/2) - Y(t, h)}{Y(t, h/4) - Y(t, h/2)}. \quad (8.25)$$

Having estimated q we can now apply Richardson's extrapolation to obtain

$$f(t) \simeq Y\left(t, \frac{h}{4}\right) + \frac{1}{2^q - 1}\left\{Y\left(t, \frac{h}{4}\right) - Y\left(t, \frac{h}{2}\right)\right\} + O(h^p). \quad (8.26)$$

Some numerical results demonstrating the effectiveness of these extrapolation techniques are given in Example 8.1.

The main advantage of Euler's method lies in its simplicity, but if extrapolation is used the method can yield reasonable accuracy with little computational work. Thus, in situations where very high accuracy is not

TABLE 8.1
Results for Example 8.1 using Euler's method and extrapolation.

s	$h=0.1$	$h=0.2$	$h=0.05$	Extrapolation using (8.21) and $h=0.1, 0.05$	Extrapolation using (8.25) and (8.26)	Correct answer
0.1		0.94849	0.95138			0.95346
0.2	0.89953	0.90768	0.91050	0.91333	0.91200	0.91288
0.3		0.87171	0.87458			0.87706
0.4	0.83336	0.83982	0.84265	0.84549	0.84487	0.84515
0.5		0.81123	0.81401			0.81650
0.6	0.77937	0.78542	0.78812	0.79082	0.79029	0.79057
0.7		0.76194	0.76456			0.76697
0.8	0.73492	0.74047	0.74301	0.74555	0.74514	0.74536
0.9		0.72074	0.72319			0.72548
1.0	0.69738	0.70251	0.70488	0.70726	0.70693	0.70711

needed, this approach may be preferable to the more accurate but also more complicated methods to be described next. Another advantage of extrapolation algorithms is that a rough estimate of the error can be obtained as part of the process.

Example 8.1. The equation

$$f(t) = \frac{1}{\sqrt{1+t}} + \frac{\pi}{8} - \frac{1}{4}\sin^{-1}\left(\frac{1-t}{1+t}\right) - \frac{1}{4}\int_0^t \frac{f(s)}{\sqrt{t-s}}\,ds,$$

has the solution $f(s) = 1/\sqrt{1+s}$.

Numerical results for several stepsizes and the values obtained by extrapolation are shown in Table 8.1.

8.3. A method based on the product trapezoidal rule. To construct higher order methods directly, it is necessary to use more accurate numerical integration rules. The next step is the *product trapezoidal method* which is constructed by approximating $K(t, s, f(s))$ by piecewise linear functions, in particular,

$$K(t, s, f(s)) \simeq \frac{s-t_j}{h} K(t, t_{j+1}, f(t_{j+1})) + \frac{t_{j+1}-s}{h} K(t, t_j, f(t_j)), \quad t_j \le s \le t_{j+1}.$$

(8.27)

This leads to the integration formula

$$\int_0^{t_n} p(t_n, s) K(t_n, s, f(s))\,ds$$

$$\simeq \alpha_{n1} K(t_n, t_0, f(t_0)) + \sum_{j=1}^{n-1} (\alpha_{n,j+1} + \beta_{nj}) K(t_n, t_j, f(t_j)) + \beta_{nn} K(t_n, t_n, f(t_n)),$$

(8.28)

where

$$\alpha_{n,j+1} = \frac{1}{h}\int_{t_j}^{t_{j+1}} (t_{j+1}-s) p(t_n, s)\,ds,$$

(8.29)

$$\beta_{n,j+1} = \frac{1}{h}\int_{t_j}^{t_{j+1}} (s-t_j) p(t_n, s)\,ds.$$

(8.30)

The numerical method for solving (8.1) is then

$$F_n = g(t_n) + \alpha_{n1} K(t_n, t_0, F_0) + \sum_{j=1}^{n-1} (\alpha_{n,j+1} + \beta_{nj}) K(t_n, t_j, F_j) + \beta_{nn} K(t_n, t_n, F_n),$$

$$n = 1, 2, \ldots, \quad (8.31)$$

with $F_0 = g(t_0)$.

If we know $F_0, F_1, \ldots, F_{n-1}$, then (8.31) is an implicit equation defining F_n, so that in principle we can solve for the approximate values in the usual stepwise process. If (8.1) is linear, then we can solve for F_n directly; however, in the nonlinear case we must be a little careful. The problem of finding F_n for nonlinear equations also arises in the case of bounded kernels, and we remarked that for small enough h we can always solve for F_n by successive substitution. For the case of bounded kernels we can take $p(x, t) = 1$ so that $\beta_{nn} = h/2$ and this iteration converges if $h |\partial K(t_n, t_n, F_n)/\partial f|/2 < 1$. For the case of unbounded kernels this situation may be much less favorable. For example, if we have $p(t, s) = 1/\sqrt{t-s}$, then $\beta_{nn} = O(\sqrt{h})$, so that convergence is now governed by the factor $\sqrt{h} |\partial K(t_n, t_n, F_n)/\partial f|$, which must be less than one, requiring often an unreasonably small h. For nonlinear problems it is therefore advisable to use a standard rootfinding method, such as the secant method or the Newton–Raphson method to solve (8.31).

8.4. A block-by-block method based on quadratic interpolation. A way of constructing highly accurate numerical methods is now fairly obvious: using product integration formulas based on approximating integrands by polynomials of degree two, three, etc., one obtains methods which are of the general form

$$F_n = g(t_n) + \sum_{j=0}^{n} w_{nj} K(t_n, t_j, F_j), \tag{8.32}$$

where expressions for the weights w_{nj} are readily found. In this way it is in principle easy to construct formulas analogous to those in § 7.2 based on the Newton–Cotes quadratures. Of course we still have the problems encountered before; several different formulas have to be combined (leading possibly to numerical instability) and starting values are required which have to be obtained in some other way. As we saw, the block-by-block methods are convenient because they eliminate these difficulties. The situation is the same here and we now describe the analogous product integration procedure. Again, we use a specific method to illustrate the general concept.

Using a piecewise quadratic interpolation polynomial, we proceed exactly as in the derivation of (7.69) and (7.70), using product integration instead of regular integration. After some manipulations, this leads to the approximating system

$$F_{2m+1} = g(t_{2m+1})$$
$$+ (1-\delta_{0m}) \sum_{i=0}^{2m} w_{2m+1,i} K(t_{2m+1}, t_i, F_i)$$
$$+ \alpha\left(t_{2m+1}, t_{2m}, \frac{h}{2}\right) K(t_{2m+1}, t_{2m}, F_{2m})$$

$$+ \beta\left(t_{2m+1}, t_{2m}, \frac{h}{2}\right) K(t_{2m+1}, t_{2m+1/2}, \tfrac{3}{8}F_{2m} + \tfrac{3}{4}F_{2m+1} - \tfrac{1}{8}F_{2m+2})$$

$$+ \gamma\left(t_{2m+1}, t_{2m}, \frac{h}{2}\right) K(t_{2m+1}, t_{2m+1}, F_{2m+1}), \tag{8.33}$$

$$F_{2m+2} = g(t_{2m+2}) + (1 - \delta_{0m}) \sum_{i=0}^{2m} w_{2m+2,i} K(t_{2m+2}, t_i, F_i)$$

$$+ \alpha(t_{2m+2}, t_{2m}, h) K(t_{2m+2}, t_{2m}, F_{2m})$$

$$+ \beta(t_{2m+2}, t_{2m}, h) K(t_{2m+2}, t_{2m+1}, F_{2m+1})$$

$$+ \gamma(t_{2m+2}, t_{2m}, h) K(t_{2m+2}, t_{2m+2}, F_{2m+2}), \tag{8.34}$$

where

$$w_{ni} = (1 - \delta_{in})\alpha(t_n, t_i, h) + (1 - \delta_{i0})\gamma(t_n, t_i - 2h, h), \quad i \text{ even}, \tag{8.35}$$

$$w_{ni} = \beta(t_n, t_i - h, h), \quad i \text{ odd}. \tag{8.36}$$

The functions α, β, γ are defined by

$$\alpha(x, y, z) = \frac{z}{2} \int_0^2 (1-s)(2-s) p(x, y + sz) \, ds, \tag{8.37}$$

$$\beta(x, y, z) = z \int_0^2 s(2-s) p(x, y + sz) \, ds, \tag{8.38}$$

$$\gamma(x, y, z) = \frac{z}{2} \int_0^2 s(s-1) p(x, y + sz) \, ds. \tag{8.39}$$

Starting with $F_0 = f(0)$, we solve (8.33) and (8.34) successively for blocks of values $(F_1, F_2), (F_3, F_4), \ldots$, using a root-finding technique such as the Newton–Raphson method in the nonlinear case.

Example 8.2. The solution computed by (8.33) and (8.34) to the equation presented in Example 8.1 for various values of h are given in Table 8.2.

We see from these results that halving the stepsize by a factor of two reduces the error by about a factor of 10. This indicates an order of convergence of

$$p \simeq \log_2 10 \simeq 3.5.$$

An intuitive explanation for this somewhat unexpected noninteger order can be given as follows. The order of convergence is the same as the order of accuracy of the product integration rule (this will be proved in the next section). Since the integration rule is based on quadratic interpolation, its accuracy is at least $O(h^3)$. Actually, if there were no singularity in the integrand, then the method would simply be Simpson's rule with accuracy $O(h^4)$. Thus near $s = t$ the integration has accuracy $O(h^3)$ while away from this point the accuracy is $O(h^4)$, and we expect the overall error to have

TABLE 8.2
Results for Example 8.1 by the quadratic block-by-block method.

s	h = 0.2	h = 0.1	h = 0.05	True answer
0.1		0.9534564	0.9534630	0.9634626
0.2	0.9128189	0.9128747	0.9128712	0.9128709
0.3		0.8770566	0.8770583	0.8770580
0.4	0.8451842	0.8451567	0.8451544	0.8451543
0.5		0.8164961	0.8164968	0.8164966
0.6	0.7905562	0.790579	0.7905695	0.7905694
0.7		0.7669648	0.7669651	0.7669650
0.8	0.7453698	0.7453570	0.7453561	0.7453560
0.9		0.7254761	0.7254763	0.7254763
1.0	0.7071018	0.7071074	0.7071068	0.7071068
$|\varepsilon_{max}| \times 10^7$	620	62	7	

order between three and four. This obviously intuitive reasoning can be made precise. For $p(t, s) = 1/\sqrt{t-s}$ one can show that the order of accuracy of the quadratic product integration is 3.5.

8.5. A convergence proof for product integration methods. We now generalize the convergence proof of Theorem 7.2 to product integration methods. For simplicity, let us consider approximation methods of the form

$$F_n = g(t_n) + \sum_{j=0}^{n} w_{nj} K(t_n, t_j, F_j), \qquad n = r, r+1, \ldots, \qquad (8.40)$$

so that, given starting values $F_0, F_1, \ldots, F_{r-1}$, we compute F_n successively, one at a time. As in § 7.3, we define the consistency error by

$$\delta(h, t_n) = \int_0^{t_n} p(t_n, s) K(t_n, s, f(s)) \, ds - \sum_{j=0}^{n} w_{nj} K(t_n, t_j, f(t_j)). \qquad (8.41)$$

We assume that $K(t, s, f(s))$ satisfies the conditions of Theorem 4.8. This is sufficient to guarantee the existence of a unique continuous solution for sufficiently small h. Subtracting (8.1) from (8.40), we get

$$\varepsilon_n = \sum_{j=0}^{n} w_{nj} \{K(t_n, t_j, F_j) - K(t_n, t_j, f(t_j))\} - \delta(h, t_n), \qquad (8.42)$$

and, since the kernel satisfies a Lipschitz condition,

$$|\varepsilon_n| \leq L \sum_{j=0}^{n} |w_{nj}| |\varepsilon_j| + |\delta(h, t_n)|. \qquad (8.43)$$

At this point we cannot apply Theorem 7.1 directly since we can no longer claim that $w_{ni} = O(h)$, as required in the proof of Theorem 7.2. To see this, consider the example discussed in § 8.2, where (8.16) indicates that $w_{n,n-1} = O(\sqrt{h})$. To proceed, we establish a somewhat complicated subsidiary result which plays essentially the same role here as did Theorem 7.1 for the case of bounded kernels.

THEOREM 8.1. *Suppose that*

$$|\varepsilon_n| \leq \sum_{j=0}^{n-1} |\alpha_{nj}||\varepsilon_j| + B, \qquad n = r, r+1, \ldots, \tag{8.44}$$

with $B > 0$ and

$$\sum_{j=0}^{r-1} |\varepsilon_j| \leq \eta.$$

(a) *If*

$$\sum_{j=0}^{n-1} |\alpha_{nj}| \leq \alpha < 1, \qquad n = r, r+1, \ldots, \tag{8.45}$$

then

$$|\varepsilon_n| \leq \frac{B+\eta}{1-\alpha}, \qquad n = 0, 1, \ldots. \tag{8.46}$$

(b) *If there exist integers $0 = j_0 < j_1 < \cdots j_m < j_{m+1}$, with $0 \leq r < j_1$ and $j_m \leq n < j_{m+1}$, such that, for $\nu = 0, 1, \ldots, m$ and $n = r, r+1, \ldots$*

$$\sum_{j=j_\nu}^{\min(n, j_{\nu+1}-1)} |\alpha_{nj}| \leq \alpha < 1, \tag{8.47}$$

then

$$|\varepsilon_n| \leq \frac{B+\eta}{(1-\alpha)^2}\left(\frac{1}{1-\alpha}\right)^m. \tag{8.48}$$

Proof. Consider the sequence p_n satisfying

$$p_n = \alpha p_{n-1} + (B+\eta), \qquad n = r, r+1, \ldots,$$
$$p_n = 0, \qquad n = 0, 1, \ldots, r-1.$$

Then clearly p_n is a nondecreasing function of n and

$$p_n \leq \frac{B+\eta}{1-\alpha}, \qquad n = 0, 1, \ldots. \tag{8.49}$$

A simple inductive argument then shows that, if (8.44) and (8.45) are satisfied, then

$$|\varepsilon_n| \leq p_n, \qquad n = r, r+1, \ldots$$

so that (8.46) follows.

To prove part (b), let
$$\xi_\nu = \max_{j_\nu \leq j \leq j_{\nu+1}} |\varepsilon_j|.$$

Then, from (8.44) and (8.47),
$$|\varepsilon_n| \leq \alpha \sum_{j=0}^{m-1} |\xi_j| + \sum_{j=j_m}^{n-1} |\alpha_{nj}| |\varepsilon_j| + B,$$

from which we get immediately
$$\xi_m \leq \frac{\alpha}{1-\alpha} \sum_{j=0}^{m-1} \xi_j + \frac{B}{1-\alpha}. \tag{8.50}$$

From (8.46) we have
$$\xi_0 \leq \frac{B+\eta}{1-\alpha},$$

and we can now apply Theorem 7.1 to get
$$\xi_m \leq \left\{ \frac{B}{1-\alpha} + \frac{\alpha(B+\eta)}{(1-\alpha)^2} \right\} \left(\frac{1}{1-\alpha} \right)^m.$$

The inequality (8.48) then follows obviously. □

This result paves the way for the convergence theorem for product integration methods.

THEOREM 8.2. *Assume that*
  (i) *the conditions of Theorem 4.5 are satisfied,*
  (ii) *the starting values $F_0, \ldots, F_{r-1}$ satisfy*
$$\sum_{j=0}^{r-1} |F_j - f(t_j)| \leq \eta,$$

(iii)
$$\lim_h \left(\max_{1 \leq n \leq N} |w_{nn}| \right) = 0,$$

(iv) *the interval $[0, T]$ can be divided into a finite number of subintervals $[0 = z_0, z_1], [z_1, z_2], \ldots, [z_{m-1}, z_m = T]$, such that if j_ν denotes the largest integer less than or equal to z_ν/h and $w_{nj} = 0$ for $j > n$, the weights w_{nj} satisfy the condition*
$$L \sum_{j=j_\nu}^{j_{\nu+1}-1} |w_{nj}| \leq \alpha < 1, \qquad n = r, r+1, \ldots, \quad \nu = 0, 1, \ldots, m. \tag{8.51}$$

This subdivision must be independent of h.
Then
  (a) *for sufficiently small h (8.40) defines a unique sequence $F_r, F_{r+1}, \ldots,$*

(b)
$$|F_n - f(t_n)| \leq \frac{|\delta(h, t_n)| + \eta}{(1-\alpha)^2}\left(\frac{1}{1-\alpha}\right)^m. \qquad (8.52)$$

Thus, if the product integration method is consistent and if the starting errors go to zero as $h \to 0$, then

$$\lim_h |\varepsilon_n| = 0. \qquad (8.53)$$

Proof. Assumption (iii) implies that $L|w_{nn}|$ can be made as small as desired by choosing a sufficiently small h. Thus, claim (a) follows from the contraction mapping theorem. For part (b), apply Theorem 8.2 to (8.43). Assumption (iv) defines the integers j_ν needed in (8.47); to get (8.52) we identify B with the consistency error $\delta(h, t_n)$. □

Conditions (iii) and (iv) on the weights w_{nj} in the above theorem are easily verifiable and hold for cases of practical interest. In fact they are closely related to and generally implied by conditions of Theorem 4.8. The proof which for the sake of simplicity was carried out only for the step-by-step methods can obviously be extended to cover the block-by-block methods. All we need do is to group values in each block into a vector $\mathbf{F}_n$, then repeat the steps of the argument using norms instead of absolute values.

Exercise 8.1. Extend Theorem 8.2 to the block-by-block method described in § 8.4.

Notes on Chapter 8. The use of product integration for approximating integrals with unbounded integrands seems to have been first suggested by Young in [248]. While the idea is relatively simple, the error analysis does involve some complications. For details, see [1] and [78].

A number of early papers ([135], [200], [233], [249]) suggest product integration for singular integral equations. Makinson and Young [176] introduced block-by-block methods in this connection. These various original ideas were later worked on and extended in [81], [82], [95], [109], [110], [158], [167].

Degenerate kernel approximations for singular equations are considered in [37] and [38].

An equation with a singularity of a type different from that discussed here is discussed in [145].

Chapter 9

Equations of the First Kind with Differentiable Kernels

As was shown in Chapter 5, equations of the first kind with smooth kernels are equivalent to equations of the second kind. Where the kernels have explicit forms which can be differentiated, one can use either (5.2) or (5.5) with the appropriate numerical method for equations of the second kind. But often explicit expressions for the kernels are not known, so that some modifications have to be made. A considerable amount of work has been done on the so-called *direct methods*, that is, methods which are based on discretizing the original equation (5.1), rather than a differentiated form. This will be our main concern in this chapter, although some comments on the use of the differentiated forms will be made. One of the unexpected results on direct method is that not all convergent numerical integration rules lead to convergent methods for the integral equation. While some of the simpler rules such as the midpoint and trapezoidal methods perform adequately, some of the well-known and more accurate rules do not.

The midpoint method produces a fairly simple and convenient direct method, but the equally simple trapezoidal method is less satisfactory. The reasons for this are discussed in § 9.2. Problems relating to the construction of higher order direct methods are outlined in §§ 9.3 and 9.4, while in § 9.5 we present some brief remarks on the solution of the equation using finite difference approximations to the derivatives. Another topic briefly considered in § 9.6 is the extension of the methods to nonlinear equations.

One difficulty with all equations of the first kind is that they are ill-posed, as was discussed in Chapter 5. This raises the possibility of a high sensitivity of the solution to small perturbations. This is investigated briefly in § 9.7.

To avoid some rather unimportant technical details, we will assume that the kernel $k(t, s)$ and the right-hand side $g(t)$ have unlimited differentiability.

For simplicity, we will consider primarily the linear case

$$\int_0^t k(t,s)f(s)\,ds = g(t), \qquad 0 \le t \le T. \tag{9.1}$$

In principle, there is no difficulty in extending the results to be given to nonlinear equations, provided some suitable conditions, such as the ones in Theorem 5.2, are satisfied.

9.1. Application of simple integration rules. The simplest approximation can be obtained by using the rectangular integration rule. The resulting method is again called *Euler's method*. Using the usual subdivision of $[0, T]$ by the points $t_i = ih$, the rectangular integration rule is

$$\int_0^{t_n} \varphi(s)\,ds \simeq h \sum_{i=0}^{n-1} \varphi(t_i), \tag{9.2}$$

which gives the discrete analogue of (9.1) as

$$h \sum_{i=0}^{n-1} k(t_n, t_i) F_i = g(t_n), \qquad n = 1, 2, \ldots. \tag{9.3}$$

Given values $F_0, F_1, \ldots, F_{n-2}$, this can be solved for F_{n-1} by

$$F_{n-1} = \frac{g(t_n)}{hk(t_n, t_{n-1})} - \sum_{i=0}^{n-2} \frac{k(t_n, t_i)}{k(t_n, t_{n-1})} F_i. \tag{9.4}$$

Since a basic assumption is that $k(t, t) \ne 0$, we know that, for sufficiently small h, $k(t_n, t_{n-1}) \ne 0$ and (9.4) will always have a solution.

Euler's method is simple, but as always, suffers from having low accuracy. As a next step we can try to use the *midpoint* integration rule which is

$$\int_0^t \varphi(s)\,ds \simeq h \sum_{i=0}^{n-1} \varphi(t_{i+1/2}), \tag{9.5}$$

where $t_{i+1/2} = t_i + h/2$. The corresponding approximation method for (9.1) is

$$h \sum_{i=0}^{n-1} k(t_n, t_{i+1/2}) F_{i+1/2} = g(t_n), \qquad n = 1, 2, \ldots. \tag{9.6}$$

The solution $F_{i+1/2}$, which approximates $f(t_{i+1/2})$, can be determined by

$$F_{n-1/2} = \frac{g(t_n)}{hk(t_n, t_{n-1/2})} - \sum_{i=0}^{n-2} \frac{k(t_n, t_{i+1/2})}{k(t_n, t_{n-1/2})} F_{i+1/2}. \tag{9.7}$$

On balance, this is perhaps the most satisfactory method for solving equations of the first kind as we will shortly see. For the moment, let us consider one more simple integration rule, namely the *trapezoidal* method. This leads to the scheme

$$\tfrac{1}{2}hk(t_n, t_n)F_n + h\sum_{i=0}^{n-1} k(t_n, t_i)F_i + \tfrac{1}{2}hk(t_n, t_0)F_0 = g(t_n) \qquad (9.8)$$

for the approximate solution. The trapezoidal method needs a starting value F_0. From (5.2) we see that

$$f(0) = \frac{g'(0)}{k(0, 0)}, \qquad (9.9)$$

which we will take for F_0.

9.2. Error analysis for simple approximation methods. As we shall see, the error analysis of methods for equations of the first kind presents some difficulties not present for equations of the second kind. It is not possible to claim that any convergent numerical integration rule leads to a convergent approximation scheme. Consequently we must take into consideration the specifics of the integration rule. We begin by analyzing the simple methods discussed in the previous section.

Consider the midpoint method (9.6). Corresponding to (9.6) we write an equation using $f(t_{i+1/2})$ instead of $F_{i+1/2}$; this gives

$$h\sum_{i=0}^{n-1} k(t_n, t_{i+1/2})f(t_{i+1/2}) = g(t_n) - \delta(h, t_n) \qquad (9.10)$$

where

$$\delta(h, t_n) = \int_0^{t_n} k(t_n, s)f(s)\, ds - h\sum_{i=0}^{n-1} k(t_n, t_{i+1/2})f(t_{i+1/2}). \qquad (9.11)$$

On subtracting (9.10) from (9.6) we get

$$h\sum_{i=0}^{n-1} k(t_n, t_{i+1/2})\varepsilon_{i+1/2} = \delta(h, t_n), \qquad (9.12)$$

where

$$\varepsilon_{i+1/2} = F_{i+1/2} - f(t_{i+1/2}). \qquad (9.13)$$

In order to obtain a bound on the error $\varepsilon_{i+1/2}$ from (9.12), we difference the equation, that is, we substract from (9.12) the same equation with n replaced by $n-1$. This is a standard approach with equations of the first kind; it is of course just the discrete analogue of converting an equation of the first kind into the more easily treated second kind form by differentiation. By this we obtain

$$hk(t_n, t_{n-1/2})\varepsilon_{n-1/2} + h\sum_{i=0}^{n-2} \{k(t_n, t_{i+1/2}) - k(t_{n-1}, t_{i+1/2})\}\varepsilon_{i+1/2}$$
$$= \delta(h, t_n) - \delta(h, t_{n-1}). \qquad (9.14)$$

From this, it is an easy matter to bound $\varepsilon_{i+1/2}$ using the following result.

THEOREM 9.1. *Assume that there exist constants c_1, c_2, and c_3, such that for all $h > 0$,*

$$\max_{\substack{1 \leq n \leq N \\ 0 \leq i \leq n-1}} |k(t_n, t_{i+1/2}) - k(t_{n-1}, t_{i+1/2})| \leq c_1 h, \quad (9.15)$$

$$\max_{1 \leq n \leq N} |\delta(h, t_n) - \delta(h, t_{n-1})| \leq c_2 h^3, \quad (9.16)$$

$$\min_{1 \leq n \leq N} |k(t_n, t_{n-1/2})| \geq c_3. \quad (9.17)$$

Then, for ε_i determined by (9.14),

$$\max_{0 \leq i \leq N-1} |\varepsilon_{i+1/2}| = O(h^2). \quad (9.18)$$

Proof. Divide (9.14) by $hk(t_n, t_{n-1/2})$, then apply Theorem 7.1. □

Exercise 9.1. Carry out a complete proof for Theorem 9.1. Establish an explicit error bound for $|\varepsilon_{i+1/2}|$.

Condition (9.15) holds if $k(t, s)$ is differentiable with respect to t, while (9.17) follows from the assumption that $k(t, t)$ is bounded away from zero. To determine when (9.16) holds we must analyze the midpoint integration rule in some detail.

From elementary results of numerical integration we know that for sufficiently smooth $\varphi(t)$

$$\int_{t_i}^{t_{i+1}} \varphi(s)\, ds = h\varphi(t_{i+1/2}) + \frac{h^3}{24} \frac{d^2}{ds^2} \varphi(s) \bigg|_{s=t_i} + O(h^5).$$

Consequently,

$$\delta(h, t_n) = \int_0^{t_n} k(t_n, s) f(s)\, ds - h \sum_{i=0}^{n-1} k(t_n, t_{i+1/2}) f(t_{i+1/2})$$

$$= \frac{h^3}{24} \sum_{i=0}^{n-1} \frac{\partial^2}{\partial s^2} k(t_n, s) f(s) \bigg|_{s=t_i} + O(h^4) \quad (9.19)$$

so that

$$\delta(h, t_n) - \delta(h, t_{n-1}) = \frac{h^3}{24} \frac{\partial^2}{\partial s^2} k(t_n, s) f(s) \bigg|_{s=t_{n-1}}$$

$$+ \frac{h^3}{24} \sum_{i=0}^{n-2} \frac{\partial^2}{\partial s^2} [k(t_n, s) f(s) - k(t_{n-1}, s) f(s)]_{s=t_i} + O(h^4). \quad (9.20)$$

If $(\partial^3/\partial t\, \partial s^2)\, k(t, s) f(s)$ is bounded, then (9.20) shows that

$$\delta_n(h, t_n) - \delta(h, t_{n-1}) = O(h^3)$$

and (9.16) is satisfied. Actually, we can say a little more: the summed term in (9.19) is just a rectangular approximation to

$$\int_0^{t_n} \frac{\partial^2}{\partial s^2} k(t_n, s) f(s) \, ds,$$

so that, if $(\partial^3/\partial s^3) k(t, s) f(s)$ is bounded, then

$$\delta(h, t_n) = \frac{h^2}{24} \frac{\partial}{\partial s} k(t_n, s) f(s) \big|_{s=0}^{s=t_n} + O(h^4).$$

We can now write (9.12) as

$$h \sum_{i=0}^{n-1} k(t_n, t_{i+1/2}) \varepsilon_{i+1/2} = \frac{h^2}{24} \frac{\partial}{\partial s} k(t_n, s) f(s) \big|_{s=0}^{s=t_n} + O(h^4), \qquad (9.21)$$

which suggests the next result.

THEOREM 9.2. *If $k(t, s)$ and $g(t)$ are sufficiently smooth, then the relation between the solution $f(t)$ of (9.1) and its approximation F_n by (9.6) is given by*

$$F_{n+1/2} = f(t_{n+1/2}) + \frac{h^2}{24} e(t_{n+1/2}) + O(h^4), \qquad (9.22)$$

where $e(t)$ is the solution of

$$\int_0^t k(t, s) e(s) \, ds = \frac{\partial}{\partial s} k(t, s) f(s) \big|_{s=0}^{s=t}. \qquad (9.23)$$

Proof. Discretize (9.23) by the midpoint method, calling the resulting approximation e_n. Then

$$h \sum_{i=0}^{n-1} k(t_n, t_{i+1/2}) e_{i+1/2} = \frac{\partial}{\partial s} k(t_n, s) f(s) \big|_{s=0}^{s=t_n}. \qquad (9.24)$$

Compare this with (9.21), difference, and apply Theorem 7.1. If all functions are sufficiently smooth so that the $O(h^4)$ term in (9.21) is a term in an expansion, then

$$\varepsilon_{i+1/2} = \frac{h^2}{24} e_{i+1/2} + O(h^4). \qquad (9.25)$$

Next, apply the steps leading to (9.14) to (9.24) and apply Theorem 9.1 to show that

$$e_{i+1/2} = e(t_{i+1/2}) + O(h^2). \qquad (9.26)$$

Putting together (9.25) and (9.26) completes the proof of (9.22). □

The result (9.22) immediately suggests the use of Richardson's extrapolation to improve the accuracy of the midpoint method. A slight change has to be made to the usual process, since halving the stepsize will result in a different set of points at which approximations to $f(t)$ are computed. The difficulty can be overcome by dividing the stepsize by three. If we let $Y(t, h)$ denote the approximate value of $f(t)$, computed with stepsize h, then

$$Y(t, h) = f(t) + \frac{h^2}{24} e(t) + O(h^4),$$

$$Y\left(t, \frac{h}{3}\right) = f(t) + \frac{h^2}{216} e(t) + O(h^4).$$

Hence

$$f(t) = Y\left(t, \frac{h}{3}\right) + \frac{1}{8}\left\{Y\left(t, \frac{h}{3}\right) - Y(t, h)\right\} + O(h^4). \tag{9.27}$$

The extrapolated values $Y^E(t)$ given by

$$Y^E(t) = Y\left(t, \frac{h}{3}\right) + \frac{1}{8}\left\{Y\left(t, \frac{h}{3}\right) - Y(t, h)\right\} \tag{9.28}$$

therefore have fourth order accuracy.

Example 9.1. The equation

$$\int_0^t \cos(t-s)f(s)\, ds = 1 - \cos t \tag{9.29}$$

has the exact solution

$$f(t) = t.$$

Results from the midpoint method and extrapolated values are shown in Table 9.1.

TABLE 9.1
Solution of (9.29) by the midpoint method and extrapolation.

t	$h = 0.3$	$h = 0.1$	Extrapolated value
0.15	0.15057	0.15006	0.14999
0.45	0.45171	0.45019	0.45000
0.75	0.75285	0.75031	0.74999
1.05	1.05398	1.05044	1.05000
1.35	1.35512	1.35056	1.34999
1.65	1.65626	1.65069	1.64999
1.95	1.95740	1.95081	1.94999

In a similar fashion one can establish convergence theorems and error estimates for Euler's method but we will not pursue this as the method is of little interest. For the trapezoidal method the convergence proof is somewhat lengthy.

THEOREM 9.3. *If $k(t, s)$ and $g(t)$ are sufficiently smooth, then the solution computed by (9.8) and (9.9) satisfies*

$$|F_n - f(t_n)| = O(h^2). \tag{9.30}$$

Proof. The arguments are somewhat technical and we refer the reader to the literature (for example, Linz [163]). □

Exercise 9.2. Show that Euler's method converges with order one.

Exercise 9.3. Prove Theorem 9.3.

While it is a fact that the solution obtained via the trapezoidal method has an accuracy of order of magnitude $O(h^2)$, no result analogous to (9.22) exists. What can be shown is that, under the appropriate assumptions on $k(t, s)$ and $g(t)$,

$$F_n = f(t_n) - \frac{h^2}{12}\{e(t_n) + (-1)^n v(t_n)\} + O(h^3), \tag{9.31}$$

where $e(t)$ is the solution of (9.23) and $v(t)$ is some other function (of no immediate interest). The conclusion which can be drawn from this is that the error in the trapezoidal rule has an oscillating component of magnitude $O(h^2)$. The results are therefore not immediately suitable for extrapolation. Actually, the oscillating component limits the usefulness of the method altogether.

Another way of looking at the difference between the midpoint and the trapezoidal methods is to consider their stability. As in Chapter 7 we can use here one of two possible approaches. The first of these involves computing the dominant term in the error expansion; for numerical stability one requires that this term satisfy an integral equation having the same kernel as the original equation. We have already done so for the midpoint method and (9.22) and (9.23) show that it is numerically stable. For the trapezoidal method we would have to make a detailed investigation of the function $v(t)$ in (9.31). Instead of doing this, we will consider the second definition of numerical stability. Here one uses a simple equation whose answer is known and whose approximation can be completely analyzed. We now take this path; it is not only easier but also more instructive.

A little thought convinces us that the appropriate equation of the first kind is

$$\int_0^t \{1 + \lambda(t-s)\} f(s)\, ds = t, \tag{9.32}$$

since after differentiating we obtain

$$f(t)+\lambda\int_0^t f(s)\,ds = 1, \qquad (9.33)$$

exactly the equation used in § 7.5. Equation (9.32) has the solution $f(t)=e^{-\lambda t}$, so with $\lambda>0$ and fixed stepsize h we expect that

$$\lim_{n\to\infty} F_n = 0 \qquad (9.34)$$

for a stable method.

Consider first the midpoint method. Putting $k(t,s)=1+\lambda(t-s)$ and $g(t)=t$ into (9.6), we get

$$h\sum_{i=0}^{n-1}\left\{1+\lambda\left(n-i-\frac{1}{2}\right)h\right\}F_{i+1/2} = hn. \qquad (9.35)$$

Differencing this once yields

$$\left(1+\frac{\lambda h}{2}\right)F_{n-1/2}+\lambda h\sum_{i=0}^{n-2} F_{i+1/2} = 1, \qquad (9.36)$$

and after a further differencing

$$\left(1+\frac{\lambda h}{2}\right)F_{n-1/2}-\left(1-\frac{\lambda h}{2}\right)F_{n-3/2} = 0. \qquad (9.37)$$

Thus

$$F_{n-1/2} = \frac{1-\lambda h/2}{1+\lambda h/2} F_{n-3/2} \qquad (9.38)$$

and (9.34) is satisfied for all $\lambda>0$, showing that the method is A-stable.

For the trapezoidal method with the above $k(t,s)$ and $g(t)$, differencing (9.8) once gives

$$\tfrac{1}{2}hF_n+\tfrac{1}{2}hF_{n-1}+\lambda h^2\sum_{j=1}^{n-1} F_j+\tfrac{1}{2}\lambda h^2 F_0 = h. \qquad (9.39)$$

Differencing once more yields

$$\tfrac{1}{2}F_{n+1}+\lambda h F_n-\tfrac{1}{2}F_{n-1} = 0. \qquad (9.40)$$

The characteristic polynomial

$$\rho^2+2\lambda h\rho-1=0$$

has a root of magnitude

$$\sqrt{1+\lambda^2 h^2}+\lambda h,$$

which is greater than one for $\lambda h > 0$. Thus, for any $\lambda h > 0$,

$$\lim_{n \to \infty} |F_n| = \infty, \tag{9.41}$$

and the trapezoidal method must be considered numerically unstable.

To summarize what we have found so far, we can say that the midpoint method is a simple and effective method for solving Volterra equations of the first kind. It has good stability properties and its accuracy can be improved by extrapolation. The trapezoidal method, although sometimes suggested in earlier references, is numerically unstable. Since in general its accuracy is lower than that of the midpoint method, there seems to be no reason at all to use it.

9.3. Difficulties with higher order methods. It is tempting to try to construct higher order methods for the solution of equations of the first kind by using some higher order Gregory or Newton–Cotes rules as was done in § 7.2 for equations of the second kind. Somewhat surprisingly, this does not work. Take for example the fourth order Gregory method and apply it to the simple equation

$$\int_0^t f(s)\,ds = 0. \tag{9.42}$$

The approximating solution is then given by

$$h \sum_{i=0}^n w_{ni} F_i = 0, \tag{9.43}$$

where the weights w_{ni} are as defined in (7.8). By differencing (9.42) we obtain

$$\tfrac{3}{8}F_n + \tfrac{19}{24}F_{n-1} - \tfrac{5}{24}F_{n-2} + \tfrac{1}{24}F_{n-3} = 0. \tag{9.44}$$

The solution of this linear difference equation is

$$c_1 \rho_1^n + c_2 \rho_2^n + c_3 \rho_3^n + c_4 \rho_4^n,$$

where c_1, c_2, c_3, c_4 depend on the starting values, and $\rho_1, \rho_2, \rho_3, \rho_4$ are the roots of the characteristic polynomial

$$\tfrac{3}{8}\rho^3 + \tfrac{19}{24}\rho^2 - \tfrac{5}{24}\rho + \tfrac{1}{24} = 0. \tag{9.45}$$

If a small initial perturbation to (9.43) is to be propagated stable, then the characteristic equation must satisfy the classical condition well known from ordinary differential equations: the roots of (9.45) must be inside or on the unit circle in the complex plane and any root of modules unity must be simple. A simple computation, however, will show that (9.45) has a root near $\rho = -2.4$, making this method nonconvergent and hence practically useless.

This difficulty is not limited to the fourth order Gregory method. It also occurs for other higher order Gregory methods as well as the Newton–Cotes formulas, such as the three-eighths and Simpson's rule. In fact, it is known that none of the standard numerical integration schemes yield convergent methods for Volterra equations of the first kind. This is one of the relatively rare instances in numerical analysis where a quite plausible approach yields a useless method.

The observation that the standard numerical integration methods are unsatisfactory has generated some interest in nonstandard rules especially designed so that they will work for equations of the first kind. The derivation of useful formulas of this type is quite lengthy, so we will sketch this topic only in some broad outlines.

Suppose that we use the three-eighths rule

$$\int_0^{3h} \varphi(s)\,ds \simeq h\{\tfrac{3}{8}\varphi(0)+\tfrac{9}{8}\varphi(h)+\tfrac{9}{8}\varphi(2h)+\tfrac{3}{8}\varphi(3h)\} \qquad (9.46)$$

in composite form to approximate an integral over $[0, t_n]$. As given, this can be done only when $n = 3m+1$. For $n = 3m+2$ and $n = 3m+3$ some adjustment has to be made in a way similar to the use of Simpson's rule in Chapter 7. Let us use the scheme depicted in Fig. 9.1, where the parameters α_1, α_2, α_3, α_4, α_5, β_1, β_2, β_3, β_4, β_5, β_6 are for the moment undetermined. One way to choose the α's and β's is to make the rule exact for polynomials up to a certain degree. For example, the first involves five parameters and could be made exact for polynomials up to degree four, while the β's could be selected to integrate polynomials of degree five exactly. However, if we did this, the result would be a combination of standard Newton–Cotes formulas and hence nonconvergent. The trick is to relax the accuracy requirement on the integration rules. If we require only that both rules are accurate for polynomials of degree three, we will be left with three undetermined parameters. We can then try to choose these so that the resulting method is convergent.

It takes a great deal of work to show that this can actually be done and to find a set of acceptable α's and β's. We will not pursue this but simply claim

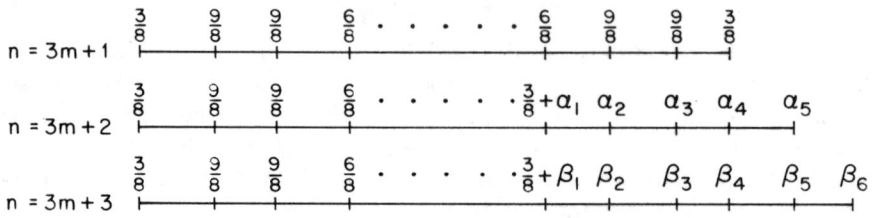

FIG. 9.1. *Weights for the modified three-eighths rule.*

that the parameters

$$\alpha_1 = 0.39740, \qquad \beta_1 = 0.43285,$$
$$\alpha_2 = 1.07706, \qquad \beta_2 = 1.02180,$$
$$\alpha_3 = 1.05107, \qquad \beta_3 = 0.95934,$$
$$\alpha_4 = 1.07706, \qquad \beta_4 = 1.24607,$$
$$\alpha_5 = 0.39740, \qquad \beta_5 = 0.87670,$$
$$\beta_6 = 0.46198,$$

give a fourth order convergent method.

A somewhat different motivation for the investigation of higher order methods comes from reconsidering the midpoint method (9.6) applied to the simple case $k(t, s) = 1$. Differencing (9.6), we see that

$$F_{n-1/2} = \frac{g(t_n) - g(t_{n-1})}{h}. \qquad (9.47)$$

The method, when solved for $F_{n-1/2}$, then reduces to a differentiation formula. Since the midpoint method is very satisfactory, one can conjecture that it is desirable for a general method to reduce to a numerical differentiation formula when used with $k(t, s) = 1$. For the general backward differentiation formula

$$g'(t_n) \simeq \frac{1}{h} \{\alpha_0 g(t_n) + \alpha_1 g(t_{n-1}) + \cdots + \alpha_k g(t_{n-k})\} \qquad (9.48)$$

we might try to choose the method so that F_n turns out to be given as

$$F_n = \frac{1}{h} \{\alpha_0 g(t_n) + \alpha_1 g(t_{n-1}) + \cdots + \alpha_k g(t_{n-k})\}. \qquad (9.49)$$

However, the F_n are determined by

$$h \sum_{i=0}^{n} w_{ni} F_i = g(t_n). \qquad (9.50)$$

If we look at this as a matrix equation

$$h\mathbf{W}\mathbf{F} = \mathbf{g} \qquad (9.51)$$

then, ignoring complications arising from possible starting values, the weight matrix W must be the inverse of the matrix A of the form

$$A = \begin{pmatrix} \ddots & & & & & & 0 \\ & \ddots & \ddots & & \ddots & & \\ & \alpha_k & \cdots & \alpha_1 & \alpha_0 & & \\ & & \alpha_k & \cdots & \alpha_1 & \alpha_0 & \\ & & & & \ddots & & \ddots \\ 0 & & & & & & \ddots \end{pmatrix}. \qquad (9.52)$$

Allowing for special starting values, it is possible to compute the weights w_{ni} for any chosen differentiation like equation (9.48). It can be shown that most backward differentiations result in convergent methods for integral equations of the first kind. The details, which are not trivial, can be found in the literature (cf. Taylor [228]).

9.4. Block-by-block methods. As we saw in the previous section, it is possible to construct higher order methods through the use of nonstandard integration rules. Unfortunately, these methods lead to formulas which are not intuitively obvious, are hard to derive, and even harder to analyze. Furthermore, they suffer from the disadvantage of requiring starting values. Many of these problems can be overcome by resorting to block-by-block methods. It turns out that the adaptation of the methods described in § 7.6 to equations of the first kind is a relatively easy matter. As in § 7.6, we describe the approach using a specific method which illustrates all the essential features of the general case.

Each interval $[t_n, t_{n+1}]$ is subdivided into three parts by the points

$$t_{n1} = t_n + \frac{h}{3}, \qquad t_{n2} = t_n + \frac{2h}{3}.$$

The approximate solution will be computed at the points t_{n1}, t_{n2}, and t_{n+1} and we will use the symbols F_{n1}, F_{n2} and F_{n+1}, respectively, to denote these approximations. To obtain these unknown values we replace the integral by a numerical integration using these points and satisfy the equation at t_{n1}, t_{n2} and t_{n+1}. For example, we may use the integration formula

$$\int_{t_n}^{t_{n+1}} \varphi(s)\, ds \simeq \frac{h}{4}\left\{3\varphi\left(t_n + \frac{h}{3}\right) + \varphi(t_{n+1})\right\}, \tag{9.53}$$

which, when scaled to the smaller intervals becomes

$$\int_{t_n}^{t_n + 2h/3} \varphi(s)\, ds \simeq \frac{h}{6}\left\{3\varphi\left(t_n + \frac{2h}{9}\right) + \varphi\left(t_n + \frac{2h}{3}\right)\right\}, \tag{9.54}$$

$$\int_{t_n}^{t_n + h/3} \varphi(s)\, ds \simeq \frac{h}{12}\left\{3\varphi\left(t_n + \frac{h}{9}\right) + \varphi\left(t_n + \frac{h}{3}\right)\right\}. \tag{9.55}$$

We want to apply these formulae to (9.1), so that $\varphi(s)$ will be identified with $k(t, s)f(s)$. Since we will have approximations to $f(t)$ only at t_{n1}, t_{n2} and t_{n+1}, everything must be expressed in terms of these values. We therefore apply to $f(t)$ a quadratic interpolation

$$f\left(t_n + \frac{h}{9}\right) \simeq \frac{1}{9}\left\{20f\left(t_n + \frac{h}{3}\right) - 16f\left(t_n + \frac{2h}{3}\right) + 5f(t_n + h)\right\}, \tag{9.56}$$

$$f\left(t_n + \frac{2h}{9}\right) \simeq \frac{1}{9}\left\{14f\left(t_n + \frac{h}{3}\right) - 7f\left(t_n + \frac{2h}{3}\right) + 2f(t_n + h)\right\}. \tag{9.57}$$

EQUATIONS OF THE FIRST KIND   155

Using these approximations in the original equation and satisfying the result at t_{n1}, t_{n2} and t_{n+1}, we are led to the system

$$\frac{k(t_{n1}, t_n + h/9)}{36}\{20F_{n1} - 16F_{n2} + 5F_{n+1}\} + \frac{k(t_{n1}, t_{n1})}{12}F_{n1}$$

$$= \frac{g(t_{n1})}{h} - \sum_{i=1}^{n-1}\{\tfrac{3}{4}k(t_{n1}, t_{i1})F_{i1} + \tfrac{1}{4}k(t_{n1}, t_{i+1})F_{i+1}\}, \quad (9.58)$$

$$\frac{k(t_{n2}, t_n + 2h/9)}{18}\{14F_{n1} - 7F_{n2} + 2F_{n+1}\} + \frac{k(t_{n2}, t_{n2})}{6}F_{n2}$$

$$= \frac{g(t_{n2})}{h} - \sum_{i=1}^{n-1}\{\tfrac{3}{4}k(t_{n2}, t_{i1})F_{i1} + \tfrac{1}{4}k(t_{n2}, t_{i+1})F_{i+1}\}, \quad (9.59)$$

$$\tfrac{3}{4}k(t_{n+1}, t_{n1})F_{n1} + \tfrac{1}{4}k(t_{n+1}, t_{n+1})F_{n+1}$$

$$= \frac{g(t_{n+1})}{h} - \sum_{i=1}^{n-1}\{\tfrac{3}{4}k(t_{n+1}, t_{i1})F_{i1} + \tfrac{1}{4}k(t_{n+1}, t_{i+1})F_{i+1}\}. \quad (9.60)$$

These three simultaneous equations are solved for $n = 1, 2, \ldots$ to yield the block of three unknowns F_{n1}, F_{n2} and F_{n+1}.

The numerical results in Example 9.2 show the effectiveness of the method.

Example 9.2. The equation

$$\int_0^t \cos(t-s)f(s)\,ds = 1 - \cos t \quad (9.61)$$

has exact solution $f(t) = t$. The errors generated by the use of equations (9.58)–(9.60) for this case are shown in Table 9.2. The results suggest that the order of convergence of the method is three.

A proof that the method is of third order is straightforward. We briefly sketch the required steps.

TABLE 9.2
Error in the approximation of (9.61) by the method of (9.58)–(9.60).

t	h = 0.4	h = 0.2	h = 0.1
0.2	—	-4.19×10^{-4}	-5.25×10^{-5}
0.4	-3.33×10^{-3}	-4.21×10^{-4}	-5.28×10^{-5}
0.6	—	-4.24×10^{-4}	-5.32×10^{-5}
0.8	-3.40×10^{-3}	-4.29×10^{-4}	-5.38×10^{-5}
1.0	—	-4.35×10^{-4}	-5.46×10^{-5}
1.2	-3.50×10^{-3}	-4.43×10^{-4}	-5.56×10^{-5}
1.4	—	-4.52×10^{-4}	-5.68×10^{-5}
1.6	-3.65×10^{-3}	-4.63×10^{-4}	-5.82×10^{-5}
1.8	—	-4.75×10^{-4}	-5.97×10^{-5}
2.0	-3.85×10^{-3}	-4.89×10^{-4}	-6.15×10^{-5}

THEOREM 9.4. *Provided $k(t, s)$ and $g(t)$ are sufficiently smooth, the approximation defined by equations* (9.58)–(9.60) *converges to the true solution of* (9.1) *with order three.*

Proof. Let $\varepsilon_{n1} = F_{n1} - f(t_{n1})$, $\varepsilon_{n2} = F_{n2} - f(t_{n2})$, and $\varepsilon_{n+1} = F_{n+1} - f(t_{n+1})$. Proceeding in the usual manner, subtract from (9.58)–(9.60) the original equations at t_{n1}, t_{n2} and t_{n+1}. This gives

$$\frac{k(t_{n1}, t_n + h/9)}{36} \{20\varepsilon_{n1} - 16\varepsilon_{n2} + 5\varepsilon_{n+1}\} + \frac{k(t_{n1}, t_{n1})}{12} \varepsilon_{n+1}$$
$$= -\sum_{i=1}^{n-1} \{\tfrac{3}{4} k(t_{n1}, t_{i1}) \varepsilon_{i1} + \tfrac{1}{4} k(t_{n1}, t_{n1}) \varepsilon_{i+1}\} + \Delta_{n1}, \quad (9.62)$$

$$\frac{k(t_{n2}, t_n + 2h/9)}{18} \{14\varepsilon_{n1} - 7\varepsilon_{n2} + 2\varepsilon_{n+1}\} + \frac{k(t_{n2}, t_{n2})}{6} \varepsilon_{n2}$$
$$= -\sum_{i=1}^{n-1} \{\tfrac{3}{4} k(t_{n2}, t_{i1}) \varepsilon_{i1} + \tfrac{1}{4} k(t_{n2}, t_{i+1}) \varepsilon_{i+1}\} + \Delta_{n2}, \quad (9.63)$$

$$\tfrac{3}{4} k(t_{n+1}, t_{n+1}) \varepsilon_{n1} + \tfrac{1}{4} k(t_{n+1}, t_{n+1}) \varepsilon_{n+1}$$
$$= -\sum_{i=1}^{n-1} \{\tfrac{3}{4} k(t_{n+1}, t_{i1}) \varepsilon_{i1} + \tfrac{1}{4} k(t_{n+1}, t_{n+1}) \varepsilon_{i+1}\} + \Delta_{n3}, \quad (9.64)$$

where the Δ_{ni} are the errors due to numerical integration and interpolation.

If we now introduce vectors

$$\boldsymbol{\varepsilon}_n = \begin{pmatrix} \varepsilon_{n1} \\ \varepsilon_{n2} \\ \varepsilon_{n+1} \end{pmatrix}, \qquad \boldsymbol{\Delta}_n = \begin{pmatrix} \Delta_{n1} \\ \Delta_{n2} \\ \Delta_{n3} \end{pmatrix},$$

then (9.62)–(9.64) can be written as

$$A_n \boldsymbol{\varepsilon}_n = \boldsymbol{\Delta}_n - \sum_{j=1}^{n-1} K_{ni} \boldsymbol{\varepsilon}_i,$$

where A_n and K_{ni} are 3×3 matrices of obvious form. From (9.62)–(9.64) we now subtract (9.64) with n replaced by $n-1$. The result can be written as

$$A_n \boldsymbol{\varepsilon}_n = \boldsymbol{\rho}_n - \sum_{j=1}^{n-1} \Delta K_{ni} \boldsymbol{\varepsilon}_i \quad (9.65)$$

where the forms of the vectors $\boldsymbol{\rho}_n$ and the matrices ΔK_{ni} are easily written down. To complete the proof we must now verify the following claims:

(a) For sufficiently small $h < h_0$ the matrices A_n have inverses which are bounded independently of h, that is,

$$\sup_{h < h_0} \|A_n^{-1}\| \leq c_1 < \infty. \quad (9.66)$$

(b) For sufficiently smooth $k(s, t)$

$$\sup_{n,i} \|\Delta K_{ni}\| \leq hc_2. \tag{9.67}$$

(c) For sufficiently smooth $k(t, s)$ and $g(t)$

$$\sup_n \|\rho_n\| \leq h^3 c_3. \tag{9.68}$$

The verification of these claims is elementary and we will leave this as an exercise. We can then multiply (9.65) by A_n^{-1} and apply Theorem 7.1. This shows that

$$\lim_h \max \|\varepsilon_n\| = 0,$$

and that $\|\varepsilon_n\| = O(h^3)$. ☐

Exercise 9.4. Verify claims (a)–(c) in Theorem 9.4. Then show that

$$\|\varepsilon_n\| = O(h^3).$$

The idea underlying the above algorithm is easily generalized. We introduce the intermediate points

$$t_n < t_{n1} < t_{n2} < \cdots < t_{n\nu} = t_{n+1},$$

then, using appropriate numerical integration and interpolation, and satisfying the equations at t_{ni}, we can write down a system of simultaneous equations for $F_{n1}, F_{n2}, \ldots, F_{n\nu}$. Convergence can then be established by repeating the proof of Theorem 9.4 in this more general context. Not only are methods of this type convergent, but it can also be shown that they are numerically stable (although the ε_{ni} have different expansions, resulting in small oscillations within each interval).

The integration method (9.53) may seem somewhat artificial; a more immediate approach would use values at both endpoints of the interval. This can be done and another class of methods is obtained by letting $t_{n1} = t_n$. However, the analysis then has to be modified since the above arguments break down. (One of the equations reduces to $0 = 0$ and has to be replaced by a requirement of continuity of the solution at t_n.) It is known that such methods are convergent, but they have a tendency to be numerically unstable and the points t_{ni} have to be selected carefully to avoid this. This potential instability should come as no surprise if we note that the trapezoidal method is a special example of this type with $\nu = 2$, $t_{n1} = t_n$, $t_{n2} = t_{n+1}$.

A generalization of the block-by-block methods can be made by using piecewise polynomial approximations to $f(t)$. The coefficients of the polynomial are determined by collocation, that is, by satisfying the resulting equations at points $t_{ni} \in [t_n, t_{n+1}]$. This involves the evaluation of moment integrals of the form

$$\int k(t,s)s^j\,ds, \qquad j=1,2,\ldots,$$

either analytically or numerically. It can then be shown that, if these integrals are evaluated by certain specific numerical methods, one obtains the block-by-block methods described above.

In connection with these methods it should be remarked that the most successful methods are those for which $t_{n1} > t_n$, so that one imposes no continuity conditions on the solution at t_n. It seems that the more smoothness one requires of the solution the worse the algorithm becomes, although it is not easy to put this observation on a rigorous basis. Using very smooth approximations such as cubic splines can lead to nonconvergent algorithms unless special care is taken.

9.5. Use of the differentiated form. If it is possible to perform analytically the differentiations for converting (9.1) into an equation of the second kind, then it is certainly appropriate to use any of the methods described in Chapter 7 to solve equations of the first kind. But even if the analytic differentiation cannot be performed, one can use numeric differentiation formulas to obtain an approximate solution. Provided the numerical differentiation is done carefully, the accuracy of the result will be affected only by a small amount. (For a fuller discussion, see § 9.7.)

The differentiated form of (9.1) is

$$k(t,t)f(t) + \int_0^t \frac{\partial}{\partial t} k(t,s)f(s)\,ds = g'(t). \tag{9.69}$$

Let us use the notation

$$D(t,s) = \frac{(\partial/\partial t)k(t,s)}{k(t,t)}, \tag{9.70}$$

and

$$G(t) = \frac{g'(t)}{k(t,t)}. \tag{9.71}$$

Then (9.69) becomes the standard equation of the second kind

$$f(t) = G(t) - \int_0^t D(t,s)f(s)\,ds. \tag{9.72}$$

If we now pick some method from Chapter 7 to compute the approximate solution by

$$F_n = G(t_n) - h \sum_{i=0}^n w_{ni} D(t_n, t_i) F_i, \qquad n = r, r+1, \ldots, \tag{9.73}$$

then, assuming no starting errors, the error will satisfy

$$|F_n - f(t_n)| = O(h^p) \tag{9.74}$$

where p is the order of the method. If $G(t)$ and $D(t, s)$ are given, nothing more has to be said. But if $G(t)$ and $D(t, s)$ can only be obtained approximately, then the effect of this error on the accuracy of the final results has to be considered.

Let us denote by $\hat{D}_{ni}$ an approximation to $D(t_n, t_i)$ and by $\hat{G}_n$ an approximation to $G(t_n)$. Let $\hat{F}_n$ be the solution computed by

$$\hat{F}_n = \hat{G}_n - h \sum_{i=0}^{n} w_{ni} \hat{D}_{ni} \hat{F}_i. \tag{9.75}$$

THEOREM 9.5. *If*

$$\max_{n,i} |D(t_{n,i}) - \hat{D}_{ni}| \leq \eta_1, \tag{9.76}$$

$$\max_{n,i} |G(t_n) - \hat{G}_n| \leq \eta_2, \tag{9.77}$$

$$\max_{n,i} |w_{ni}| \leq W, \tag{9.78}$$

and if $F_i = \hat{F}_i$, for $i = 0, 1, \ldots, r-1$, then

$$|\hat{F}_n - F_n| = O(\eta_1) + O(\eta_2). \tag{9.79}$$

Proof. Setting

$$\hat{\varepsilon}_i = \hat{F}_i - F_i,$$

and subtracting (9.73) from (9.75), we get

$$\hat{\varepsilon}_n = \hat{G}_n - G_n - h \sum_{i=0}^{n} w_{ni} \hat{D}_{ni} \hat{\varepsilon}_i + h \sum_{i=0}^{n} w_{ni} (D_{ni} - \hat{D}_{ni}) F_i.$$

From this it follows that

$$|\hat{\varepsilon}_n| \leq \eta_2 + hWD \sum_{i=0}^{n} |\hat{\varepsilon}_i| + \eta_3,$$

where $D = \max |D_{ni}|$, and $\eta_3 = O(\eta_1)$.

The result (9.79) then follows by applying Theorem 7.1. Since by assumption

$$|F_n - f(t_n)| = O(h^p),$$

we see that

$$|\hat{F}_n - f(t_n)| = O(h^p) + O(\eta_1) + O(\eta_2). \qquad \square$$

Exercise 9.5. Improve the order of magnitude estimate (9.79) by establishing an explicit bound for $|\hat{F}_n - F_n|$ in terms of η_1, η_2 and $\max |F_i|$.

The not unexpected conclusion then is that, in order to preserve the order of convergence p of the method, the numerical differentiation formulas for approximating $\partial/\partial t\, k(t, s)$ and $g(t)$ should also have order of accuracy p.

Not much has been said about this approach in the literature on the numerical solution of Volterra integral equations of the first kind. The reason is probably that it offers little apparent advantage over the direct methods. Due to the necessity of computing numerical derivatives, the formulas are more complicated and the computations somewhat more time consuming. A direct approach via a block-by-block method is in general easier.

9.6. Nonlinear equations. Although we have concentrated on the linear case, the methods described can be extended to nonlinear problems. For the equation

$$\int_0^t K(t, s, f(s))\, ds = g(t), \qquad (9.80)$$

equations (9.58) to (9.60) become

$$\tfrac{1}{4} K\!\left(t_{n1}, t_n + \tfrac{h}{9}, \tfrac{20}{9} F_{n1} - \tfrac{16}{9} F_{n2} + \tfrac{5}{9} F_{n+1}\right) + \tfrac{1}{12} K(t_{n1}, t_{n1}, F_{n1})$$

$$= \frac{g(t_{n1})}{h} - \sum_{i=1}^{n-1} \{\tfrac{3}{4} K(t_{n1}, t_{i1}, F_{i1}) + \tfrac{1}{4} K(t_{n1}, t_{i+1}, F_{i+1})\}, \qquad (9.81)$$

$$\tfrac{1}{2} K\!\left(t_{n2}, t_n + \tfrac{2h}{9}, \tfrac{14}{9} F_{n1} - \tfrac{7}{9} F_{n2} + \tfrac{2}{9} F_{n+1}\right) + \tfrac{1}{6} K(t_{n2}, t_{n2}, F_{n2})$$

$$= \frac{g(t_{n2})}{h} - \sum_{i=1}^{n-1} \{\tfrac{3}{4} K(t_{n2}, t_{i1}, F_{i1}) + \tfrac{1}{4} K(t_{n1}, t_{i+1}, F_{i+1})\}, \qquad (9.82)$$

$$\tfrac{3}{4} K(t_{n+1}, t_{n1}, F_{n1}) + \tfrac{1}{4} K(t_{n+1}, t_{n+1}, F_{n+1})$$

$$= \frac{g(t_{n+1})}{h} - \sum_{i=1}^{n-1} \{\tfrac{3}{4} K(t_{n+1}, t_{i1}, F_{i1}) + \tfrac{1}{4} K(t_{n+1}, t_{i+1}, F_{i+1})\}. \qquad (9.83)$$

In principle, these three simultaneous nonlinear equations can be solved for the unknowns, F_{n1}, F_{n2}, and F_{n+1} by standard numerical techniques, such as Newton's method. The only difficulty here is to show that the system (9.81)–(9.83) does in fact have a unique solution. This is a somewhat tedious matter which can be attacked using known results for nonlinear equations, in particular, the Kantorovich theorem which can be found in books on advanced numerical analysis. The conclusion is that, if the conditions of Theorem 5.2 are satisfied, then the system (9.81)–(9.83) has a unique solution for sufficiently small h.

9.7. Some practical considerations. Although there is a variety of methods for the numerical solution of Volterra integral equations of the first kind, any given practical problem may impose some limitations or requirements which might make one or the other of these methods more suitable than the rest. We now briefly sum up our views on the question of method selection in practice.

If we can explicitly perform the differentiations required to get (9.69) and the resulting formulas are convenient for numerical work, the best method is likely to be the numerical solution of the equivalent equation of the second kind. If explicit differentiation cannot be performed, but $\partial k(t, s)/\partial t$ and $g'(t)$ can be tabulated to adequate accuracy, it may still be reasonable to solve (9.69) instead of (9.1).

If we wish to tackle the first kind equation (9.1) directly, the midpoint method is quite convenient and together with extrapolation can give good accuracy. The trapezoidal method seems less satisfactory, since it is numerically unstable. When high accuracy is required, the block-by-block methods may be most suitable. Some of the more recently developed methods, such as those based on piecewise polynomial or splines approximations or step-by-step methods using nonstandard quadratures, look promising, but no systematic study has been done to show that the increased complexity is compensated by additional efficiency.

Methods based on (9.73) have received little attention. They appear to be somewhat unwieldy, since they require numerical differentiation. When high order differences are used, this becomes cumbersome near the ends of the interval of integration. It is not clear whether such methods have any advantage at all.

There is one point not dealt with so far that merits some attention. In proving convergence for the various methods we have ignored the possible presence of errors other than those arising from the replacement of the integral by a sum. Additional errors are, however, usually present. There is of course always some round-off error whenever numerical computations are performed. Also, if equation (9.1) arises from an experimental situation, then observational errors will affect the solution. Therefore, we must consider the effect of small perturbations on the computed solution.

One reason why this point needs to be of concern is that in some sense (9.1) is not a well-posed problem. Without making a precise definition of this term, we can say that an equation is well-posed if a small change in the parameters of the problem can cause only a small change in the solution. Now it is possible to perturb $g(t)$ in (9.1) by a small amount in such a way as to change $g'(t)$ by an arbitrarily large amount. Thus, because of the equivalence between (9.1) and (9.69) a small change in g can cause a very large change in f and the problem is not well-posed.

Problems which are not well-posed are generally difficult to solve numerically since small errors can be magnified significantly. The classical case of an ill-posed problem is the Fredholm integral equation of the first kind. The solution of such an equation calls for special methods, usually involving some form of *regularizing* or smoothing the solution. Several authors have recognized the difficulty and proposed the use of some classical regularization techniques. However, the problem is not nearly as serious for Volterra equations as it is for the Fredholm case. While error magnification does occur, it is possible to keep it more or less under control.

To understand this a little better, let us take the midpoint method as an example. This simplifies the analysis and illustrates an effect common to the other methods. Reconsider (9.6), perturbing the right-hand side by an amount η_n. The corresponding solution $\hat{F}_{n-1/2}$ is then determined by

$$h \sum_{i=0}^{n-1} k(t_n, t_{i+1/2}) \hat{F}_{i+1/2} = g(t_n) + \eta_n. \tag{9.84}$$

THEOREM 9.6. *Let $F_{n-1/2}$ and $\hat{F}_{n-1/2}$ denote the respective solutions of (9.6) and (9.84). If*

$$\max_{1 \leq n \leq N} |\eta_n| \leq \eta, \tag{9.85}$$

and if the conditions of Theorem 9.2 are satisfied, then, for $h \leq h_0$,

$$|F_{n-1/2} - \hat{F}_{n-1/2}| \leq \frac{2\eta}{Eh} \exp\left(\frac{Ct_n}{E}\right), \tag{9.86}$$

where

$$C = \max_{0 \leq s \leq t \leq T} \left|\frac{\partial}{\partial t} k(t, s)\right|, \tag{9.87}$$

and

$$E = \min_{\substack{0 \leq t \leq T \\ 0 \leq h \leq h_0}} \left|k\left(t, t - \frac{h}{2}\right)\right|. \tag{9.88}$$

Proof. Subtract (9.6) from (9.84). Then

$$h \sum_{i=0}^{n-1} |k(t_n, t_{i+1/2})| |F_{i+1/2} - \hat{F}_{i+1/2}| \leq \eta. \tag{9.89}$$

Differencing this equation and dividing by $|k(t_n, t_{n-1/2})|$ gives

$$|F_{n-1/2} - \hat{F}_{n-1/2}| \leq \frac{hC}{E} \sum_{i=0}^{n-2} |F_{i+1/2} - \hat{F}_{i+1/2}| + \frac{2\eta}{hE}. \tag{9.90}$$

Applying Theorem 7.1 and inequality (7.18), then completes the argument. □

By taking the special case $k(s, t) = 1$ and $\eta_n = (-1)^n \eta$ we see that the bound (9.86) can in fact be attained, that is, the error is magnified by a factor $2/h$. This error magnification is inherent in the problem and is not due to any defect of the midpoint method. Any other algorithm we might try leads to similar (or perhaps worse) results. For example, if we approach the problem via (9.69), then numerical differentiation will magnify the errors by a factor of $O(1/h)$, thus changing the solution by a similar amount. Thus whatever method we use there will be error magnification; fortunately it is of a limited extent and generally manageable.

From (9.86) we then draw some general conclusions. Unless very high accuracy is required, round-off error is not likely to present a serious problem. On a typical computer the error due to round-off might be of order 10^{-8} or 10^{-9}. Practical limitations make it unlikely that one would use a stepsize much smaller than 10^{-3}, so that we should be able to obtain an accuracy of at least 10^{-5}. This perhaps also explains why the difficulty has largely been ignored by most writers on this subject. The numerical examples usually show an accuracy several orders of magnitude below the machine accuracy and round-off is therefore not significant.

The effect of experimental errors may be more serious. If an error is about 1% of the true value, then for $h = 10^{-2}$ the effect on the solution may be of the same order of magnitude as the solution itself, completely invalidating the results. In this situation considerable care must be taken. One precaution would be to smooth all data which is likely to be contaminated with experimental errors. Whether this by itself is sufficient is debatable. So far, this aspect of Volterra equations of the first kind has received little attention. What the best methods are when large experimental errors are present appears to be an open problem.

Notes on Chapter 9. Although some of the older references [69], [87] make brief mention of numerical methods, no systematic study was made until Jones [146] studied the trapezoidal method for convolution equations. The start of the development of a general theory began with the work of Kobayashi [153] and Linz [163]. The trapezoidal method is analyzed in detail in [153], while in [163], [165] the midpoint method is developed. The use of product integration in connection with some simple methods is explored in [3] and [168].

The fact that many of the higher order integration rules lead to nonconvergent methods was pointed out in [163]. Later, Gladwin and Jeltsch [114] showed that this was true for all standard interpolatory integration rules. No higher order direct methods were known until de Hoog and Weiss [79], [80] produced an analysis for block-by-block methods. Brunner [43], [49], [50], [52], [53], [55], [56] analyzed closely related collocation methods. A study of spline approximations can be found in [94].

A thorough analysis of nonstandard modified Newton–Cotes formulas is given in [128]. The numbers in § 9.3 were taken from this reference. Taylor [228] studied the relation between methods for first kind equations and differentiation. Further investigations of this and related ideas are in [12], [113], [115], [129], [145], [183].

The difficulty arising from the presence of large errors and their effect on standard direct methods is pointed out in [169]. Some authors [87], [130], [219] also recognize this problem and suggest the use of classical regularization. However, the work of Radziuk [214] on least squares methods give an indication that regularization may sometimes be unnecessary. Radziuk's empirical observations are given some theoretical support in [170].

Chapter 10

Equations of the Abel Type

A logical continuation of the development of the previous chapters would be to consider now equations of the first kind with nondifferentiable or unbounded kernels. The completely general case is quite difficult and virtually nothing has been done on it. Fortunately, practical situations almost always lead to cases in which the singularity is of the form $(t-s)^{-\mu}$, $0 < \mu < 1$. We therefore consider the generalized Abel equation

$$\int_0^t (t-s)^{-\mu} k(t, s) f(s) \, ds = g(t), \qquad (10.1)$$

where $0 < \mu < 1$ and $k(t, s)$ is a smooth function. To assure the existence of a solution, we make the assumption that $k(t, t) \neq 0$ for all value of t.

Of considerable practical interest is the simple Abel equation for which $k(t, s) = 1$. The explicit inversion formula discussed in Chapter 5 permits a straightforward numerical approach using product integration. This is discussed in § 10.1. For the more general case (10.1) an immediate discretization, similar to the techniques used in Chapter 7, is readily made. While the formulas are easily derived, their analysis has proved to be a difficult matter. Of all the numerical techniques described in this book, this is the least understood. A number of isolated results, often technically difficult, have been made, but at this point no comprehensive theory exists. In §§ 10.2 to 10.4 we discuss some of the methods and provide some introduction to the less complicated aspects of the analysis.

Abel equations arise most frequently in reconstruction and inference problems, as indicated by the examples in § 2.4. Thus the right-hand side of (10.1) may be known only within an experimental inaccuracy. Since Abel's equation is somewhat ill-posed, this matter must be considered carefully. Section 10.5 is devoted to some of the practical aspects of solving Abel's equation.

10.1. Solving a simple Abel equation. As pointed out in § 5.2, the simple Abel equation

$$\int_0^t (t-s)^{-\mu} f(s)\, ds = g(t) \tag{10.2}$$

has the solution

$$f(t) = \frac{\sin \mu\pi}{\pi} \frac{d}{dt} \int_0^t (t-s)^{\mu-1} g(s)\, ds, \tag{10.3}$$

or, with $g(0) = 0$,

$$f(t) = \frac{\sin \mu\pi}{\pi} \int_0^t (t-s)^{\mu-1} g'(s)\, ds. \tag{10.4}$$

Using numerical differentiation and product integration, the approximate solution of (10.3) and (10.4) is an elementary matter. For example, one might use the differentiation formula

$$\varphi'(t) \simeq \frac{1}{2h}\{-\varphi(t_{2i+2}) + 4\varphi(t_{2i+1}) - 3\varphi(t_{2i})\}$$

$$+ \frac{t - t_{2i}}{h^2}\{\varphi(t_{2i+2}) - 2\varphi(t_{2i+1}) + \varphi(t_{2i})\}, \tag{10.5}$$

which for $\varphi \in C^{(3)}[t_{2i}, t_{2i+2}]$ and $t_{2i} \leq t \leq t_{2i+2}$ is an approximation with error $O(h^2)$. One can then either apply (10.5) to (10.4) with $\varphi = g$ and use product integration to evaluate f, or alternately use product integration first to evaluate the integral in (10.3) approximately, then set

$$\varphi(t) = \int_0^t (t-s)^{\mu-1} g(s)\, ds$$

in (10.5).

This is straightforward, but some caution has to be used. Both alternatives involve numerical differentiation, a process which may cause error magnification. As is obvious from (10.5), a perturbation in $\varphi(t)$ of order ε can cause an error in the approximate derivative of order ε/h. Thus, this simple approach is useful only when an error of magnitude $O(\varepsilon/h)$ is within acceptable limits.

10.2. The midpoint and trapezoidal methods for general Abel equations. Equations (5.38) and (5.39) show that the general Abel equation (10.1) can be converted into an integral equation with a bounded and smooth kernel. However, there seems to be little practical usefulness in this observation, since the kernel in (5.39) is rather complicated. A more direct approach, using a discretization of (10.1) and product integration is much easier.

If for $t_i \leq s \leq t_{i+1}$, we approximate $k(t,s)f(s)$ by a constant function $k(t, t_{i+1/2})f(t_{i+1/2})$ and use product integration to evaluate the integral, we get the approximation method

$$\sum_{i=0}^{n-1} w_{ni} k(t_n, t_{i+1/2}) F_{i+1/2} = g(t_n), \quad n = 1, 2, \ldots \quad (10.6)$$

where

$$w_{ni} = \int_{t_i}^{t_{i+1}} (t_n - s)^{-\mu} \, ds. \quad (10.7)$$

This is the analogue of the midpoint method (9.6). The solution can be computed step by step via

$$F_{n-1/2} = \frac{g(t_n)}{w_{n,n-1} k(t_n, t_{n-1/2})} - \sum_{i=0}^{n-2} \frac{w_{ni}}{w_{n,n-1}} \frac{k(t_n, t_{i+1/2})}{k(t_n, t_{n-1/2})} F_{i+1/2}. \quad (10.8)$$

Since by assumption $k(t, t) \neq 0$, equation (10.8) has a solution for sufficiently small h.

The midpoint method for Abel's equation is simple, but its behavior is no longer as satisfactory as it is for equations with smooth kernels.

Example 10.1. Numerical solutions for the trial equation

$$\int_0^t \frac{(1+s)f(s)}{\sqrt{t-s}} \, ds = \tfrac{16}{15} t^{5/2} + \tfrac{4}{3} t^{3/2}, \quad (10.9)$$

using the midpoint method with several values of h are shown in Table 10.1. To estimate the order of convergence and perhaps improve the accuracy of the solution, Aitken's method, described in Chapter 8, was tried. Allowing for the fact that in the midpoint method the stepsize has to be reduced by a factor of 3 to allow extrapolation, we find that the order of convergence p can be estimated by

$$p = \log_3 \left\{ \frac{Y(t, h/3) - Y(t, h)}{Y(t, h/9) - Y(t, h/3)} \right\},$$

where $Y(t, h)$ is the approximation at t, using stepsize h.

TABLE 10.1

Approximations for (10.9) by the midpoint method, with estimated order of convergence p and extrapolated values Y^E. The exact solution of (10.9) is $f(t) = t$.

t	$h = \tfrac{1}{10}$	$h = \tfrac{1}{30}$	$h = \tfrac{1}{90}$	p	Y^E
0.05	0.0686	0.0541	0.0508	1.35	0.0498
0.15	0.1643	0.1528	0.1505	1.46	0.1499
0.25	0.2627	0.2525	0.2505	1.48	0.2500
0.35	0.3619	0.3523	0.3505	1.52	0.3501

Once p is estimated, an extrapolated value Y^E is computed by

$$Y^E = \frac{3^p Y(t, h/9) - Y(t, h/3)}{3^p - 1}.$$

The computed values of p in Table 10.1 strongly suggest that the method has (at least for this example) an order of convergence $\frac{3}{2}$. The extrapolated values have improved accuracy.

To develop more accurate methods requires a better approximation to $k(t, s)f(s)$. Actually, we have already discussed this in Chapter 8 in connection with equations of the second kind. All we need to do is to adapt the relevant formulas to Abel equations taking into account that we are dealing with equations of the first kind. For example, using the results in § 8.3 on the product trapezoidal method, we find that an appropriate approximating equation for (10.1) is

$$\sum_{j=0}^{n} w_{nj} k(t_n, t_j) F_j = g(t_n), \qquad n = 1, 2, \ldots \qquad (10.10)$$

with

$$\begin{aligned} w_{n0} &= \alpha_{n1}, \\ w_{nj} &= \alpha_{n,j+1} + \beta_{nj}, \qquad j = 1, 2, \ldots, n-1, \\ w_{nn} &= \beta_{nn}, \end{aligned} \qquad (10.11)$$

where the α's and β's are given by (8.29) and (8.30).

This method requires a starting value F_0. One possible way to get this is to consider (10.1) in the limit as t approaches zero. Suppose that near $t = 0$

$$g(t) = \{C + o(1)\} t^{-\alpha}. \qquad (10.12)$$

Then, as indicated in the proof of Theorem 5.3,

$$\frac{d}{dx} \int_0^x \frac{g(t)\,dt}{(x-t)^\mu} \simeq C \frac{\Gamma(\mu)\Gamma(1-\alpha)}{\Gamma(1+\mu-\alpha)(\mu-\alpha)} x^{\mu-\alpha-1}.$$

From (5.39)

$$h(x, x) = k(x, x) \int_0^1 \frac{du}{u^\mu (1-\mu)^{1-\mu}}\,du = \Gamma(\mu)\Gamma(1-\mu) k(x, x). \qquad (10.13)$$

Differentiating (5.38), using (10.12) and (10.13), and letting $x \to 0$, we see that

$$f(0) = \begin{cases} 0 & \text{if } \mu - \alpha > 1, \\ \dfrac{C\Gamma(1-\alpha)}{\Gamma(1-\mu) k(0, 0)} & \text{if } \mu - \alpha = 1. \end{cases}$$

When $\mu - \alpha < 1$, the solution is unbounded at $t = 0$. In this case different techniques must be used; for example, we can introduce a new function

EQUATIONS OF THE ABEL TYPE

TABLE 10.2
Approximate solution of (10.15) by the method (10.10).

t	$h = 0.2$	$h = 0.1$	$h = 0.05$	Exact solution
0.2	0.832478	0.833129	0.833282	0.833333
0.4	0.713418	0.714066	0.714230	0.714286
0.6	0.624219	0.624804	0.624951	0.625000
0.8	0.554890	0.555388	0.555514	0.555556
1.0	0.499432	0.499858	0.499964	0.500000

$\varphi(t) = t^\beta f(t)$ such that $\varphi(t)$ is bounded. The modified equation can then be treated by product integration; this is fairly straightforward and we will not pursue it.

If $f(0)$ is bounded, we simply set $F_0 = f(0)$ and rearrange (10.10) to

$$F_n = \frac{1}{w_{nn} k(t_n, t_n)} \left\{ g(t_n) - \sum_{j=0}^{n-1} w_{nj} k(t_n, t_j) F_j \right\}, \quad n = 1, 2, \ldots, \tag{10.14}$$

which can be used to obtain successive values of F_n.

Example 10.2. The equation

$$\int_0^t \frac{(1+t+s) f(t)}{\sqrt{t-s}} = \frac{2t}{\sqrt{t+1}} \tanh^{-1}\left(\sqrt{\frac{t}{t+1}}\right) + 2\sqrt{t} \tag{10.15}$$

has the exact solution

$$f(t) = \frac{1}{1+t}. \tag{10.16}$$

Approximate values computed by (10.10) for several stepsizes are shown in Table 10.2. The observed error indicates a convergence of order two.

10.3. Block-by-block methods. Working along the lines indicated in the previous section, we can find any number of alternative formulae by using various product integration rules. The idea behind the block-by-block methods can also be used to obtain higher order self-starting methods. Thus, the method described in § 8.4 can immediately be adapted for the generalized Abel equation. If in (8.33) and (8.34) we omit the term on the left-hand side and rewrite the rest, making use of the fact that our equation is linear, we obtain an approximation defined by

$$A_m \begin{pmatrix} F_{2m+1} \\ F_{2m+2} \end{pmatrix} = \begin{pmatrix} S_{m1} \\ S_{m2} \end{pmatrix}, \tag{10.17}$$

where A_m is a 2×2 matrix with elements

$$A_m(1, 1) = \tfrac{3}{4}\beta\left(t_{2m+1}, t_{2m}, \frac{h}{2}\right)k(t_{2m+1}, t_{2m+1/2})$$

$$+ \gamma\left(t_{2m+1}, t_{2m}, \frac{h}{2}\right)k(t_{2m+1}, t_{2m+1}), \quad (10.18)$$

$$A_m(1, 2) = -\tfrac{1}{8}\beta\left(t_{2m+1}, t_{2m}, \frac{h}{2}\right)k(t_{2m+1}, t_{2m+1/2}), \quad (10.19)$$

$$A_m(2, 1) = \beta(t_{2m+2}, t_{2m}, h)k(t_{2m+2}, t_{2m+1}), \quad (10.20)$$

$$A_m(2, 2) = \gamma(t_{2m+2}, t_{2m}, h)k(t_{2m+2}, t_{2m+2}). \quad (10.21)$$

The values on the right-hand side of (10.17) are given by

$$S_{m1} = g(t_{2m+1}) - (1-\delta_{0m}) \sum_{i=0}^{2m} w_{2m+1,i} k(t_{2m+1}, t_i) F_i$$

$$- \left\{\alpha\left(t_{2m+1}, t_{2m}, \frac{h}{2}\right)k(t_{2m+1}, t_{2m})\right.$$

$$\left. + \tfrac{3}{8}\beta\left(t_{2m+1}, t_{2m}, \frac{h}{2}\right)k(t_{2m+1}, t_{2m+1/2})\right\} F_{2m}, \quad (10.22)$$

$$S_{m2} = g(t_{2m+2}) - (1-\delta_{0m}) \sum_{i=0}^{2m} w_{2m+2,i} k(t_{2m+2}, t_i) F_i$$

$$- \alpha(t_{2m+2}, t_{2m}, h)k(t_{2m+2}, t_{2m}) F_{2m}. \quad (10.23)$$

The w_i, α, β, γ are as defined in (8.35)–(8.39) with $p(t, s) = (t-s)^{-\mu}$.

TABLE 10.3
Observed errors for the solution of (10.24) by method (10.15).

t	$h = 0.2$	$h = 0.1$	$h = 0.05$
0.1		−0.000158	0.000031
0.2	−0.000935	0.000212	0.000029
0.3		−0.000098	0.000023
0.4	0.001231	0.000159	0.000018
0.5		−0.000057	0.000014
0.6	−0.000404	0.000102	0.000011
0.7		−0.000033	0.000008
0.8	0.000663	0.000065	0.000007
0.9		−0.000020	0.000005
1.0	−0.000169	0.000042	0.000004

Example 10.3. The method of (10.17) was used on

$$\int_0^t \frac{(1+t+s)f(s)}{\sqrt{t-s}} \, ds = \frac{2t}{\sqrt{t+1}} \tanh^{-1}\left(\sqrt{\frac{t}{t+1}}\right) + 2\sqrt{t}, \quad (10.24)$$

whose solution is

$$f(t) = \frac{1}{1+t}. \quad (10.25)$$

The observed errors for various stepsizes are shown in Table 10.3.

10.4. Some remarks on error analysis. While it is easy to think of any number of plausible methods for the approximate solution of Abel equations, considerable difficulty arises when we try to analyze the results. The techniques used in Chapter 9 for equations with smooth kernels are not easily modified for this case. The difficulty arises because the weights w_{nj} no longer follow a simple pattern, so that differencing does not produce an equation to which the basic Theorem 7.1 can be applied. The recurrence relations which do arise must be analyzed using methods which are extremely complex. Several approaches have been tried, but a clear-cut preferred choice has not yet emerged. We will briefly outline here a method (without rigorously justifying it) which seems to hold some promise.

To avoid as many technical difficulties as possible, let us take the simplest case, namely equation (10.1) with $k(t,s)=1$ and $\mu=\frac{1}{2}$, using the midpoint method (10.6). The approximate solution is then given by

$$F_{n-1/2} = \frac{g(t_n)}{w_{n,n-1}} - \sum_{i=0}^{n-2} \frac{w_{ni}}{w_{n,n-1}} F_{i+1/2} \quad (10.26)$$

where

$$w_{ni} = 2\{\sqrt{t_n - t_i} - \sqrt{t_n - t_{i+1}}\}, \quad i = 0, 1, \ldots, n-1 \quad (10.27)$$
$$= 2\sqrt{h}\{\sqrt{n-i} - \sqrt{n-i-1}\}. \quad (10.28)$$

By the usual arguments one can deduce that the error $\varepsilon_{n-1/2} = F_{n-1/2} - f(t_{n-1/2})$ satisfies

$$\sum_{i=0}^{n-1} w_{ni}\varepsilon_{i+1/2} = \delta(h, t_n), \quad (10.29)$$

where $\delta(h, t_n)$ is the error caused by the use of the product midpoint method. From this we get

$$\varepsilon_{n-1/2} = \frac{\delta(h, t_n)}{w_{n,n-1}} - \sum_{i=0}^{n-2} \frac{w_{ni}}{w_{n,n-1}} \varepsilon_{i+1/2}, \quad n = 1, 2, \ldots. \quad (10.30)$$

The hard question is to obtain a bound on the solution of this recurrence relation. Since there is no factor h in front of the summed term, establishing such a bound requires a detailed investigation of the weights w_{ni}.

First, notice that w_{ni} is a function only of $n-i$. We can therefore write (10.30) as

$$\begin{aligned} a_0 \varepsilon_{1/2} &= \Delta_1, \\ a_1 \varepsilon_{1/2} + a_0 \varepsilon_{3/2} &= \Delta_2, \\ a_2 \varepsilon_{1/2} + a_1 \varepsilon_{3/2} + a_0 \varepsilon_{5/2} &= \Delta_3 \\ &\vdots \end{aligned} \qquad (10.31)$$

where

$$\Delta_n = \frac{\delta(h, t_n)}{w_{n,n-1}}, \qquad (10.32)$$

and

$$a_j = \frac{w_{n,n-j-1}}{w_{n,n-1}}. \qquad (10.33)$$

In matrix notation we write this as

$$A\varepsilon = \Delta \qquad (10.34)$$

and the analysis then hinges on being able to find a bound for the inverse of the matrix A.

The matrix A has a rather special form and is called a *semicirculant* matrix. Not only is it triangular, but its coefficients depend only on the difference between the row and column numbers. There exist some classical results which give information on the inverses of such matrices. If we let $B = A^{-1}$ then B is also a semicirculant matrix, that is,

$$B = \begin{pmatrix} b_0 & & & & \\ b_1 & b_0 & & & \\ b_2 & b_1 & b_0 & & \\ \vdots & & \ddots & \ddots & \\ \vdots & & \cdots & b_1 & b_0 \end{pmatrix}, \qquad (10.35)$$

where the b_i can be obtained by the step-by-step formula

$$\begin{aligned} a_0 b_0 &= 1, \\ a_1 b_0 + a_0 b_1 &= 0, \\ a_2 b_0 + a_1 b_1 + a_0 b_2 &= 0, \\ &\vdots \\ a_n b_0 + a_{n-1} b_1 + \cdots + a_0 b_n &= 0. \end{aligned} \qquad (10.36)$$

If we now consider the two infinite series

$$\sum_{i=0}^{\infty} a_i x^i \quad \text{and} \quad \sum_{i=0}^{\infty} b_i x^i,$$

then, at least formally, we can consider (10.36) to be equivalent to

$$\sum_{i=0}^{\infty} a_i x^i \sum_{i=0}^{\infty} b_i x^i = 1. \tag{10.37}$$

To see this, expand (10.37) and equate powers of x. This yields (10.36). We further rewrite this as

$$\sum_{i=0}^{\infty} b_i x^i = \left\{ \sum_{i=0}^{\infty} a_i x^i \right\}^{-1}. \tag{10.38}$$

To proceed we must now use a theorem dealing with the inverse of certain infinite series.

THEOREM 10.1. *If*

$$p(x) = \sum_{i=0}^{\infty} p_n x^n \tag{10.39}$$

is convergent for $|x| < 1$, *with* $p_n > 0$, $p_0 = 1$, *and*

$$\frac{p_{n+1}}{p_n} \geq \frac{p_n}{p_{n-1}}, \tag{10.40}$$

then

$$\{p(x)\}^{-1} = 1 - \sum_{i=1}^{\infty} c_i x^i,$$

with

$$c_i \geq 0 \quad \text{and} \quad \sum_{i=1}^{\infty} c_i \leq 1. \tag{10.41}$$

Proof. A proof of this result can be found in Hardy [122, p. 68].

To apply Theorem 10.1 to (10.31), we must show that the a_j defined by (10.33) satisfy (10.40). This is a relatively easy matter. The theorem then guarantees that $b_n \geq 0$ and

$$\sum_{i=1}^{\infty} b_i \leq 1. \tag{10.42}$$

From (10.34) and the fact that $B = A^{-1}$ we see that

$$\varepsilon_{n-1/2} = \sum_{i=1}^{n} b_{n-i} \Delta_i. \tag{10.43}$$

Therefore

$$|\varepsilon_{n-1/2}| \leq \max_{1 \leq i \leq n} |\Delta_i| \sum_{i=0}^{\infty} b_i \leq 2 \max_{1 \leq i \leq n} |\Delta_i|. \tag{10.44}$$

The product midpoint method is based on a piecewise constant approximation. It follows therefore easily from the discussion of § 8.1 that $\delta(h, t_n)$ is at least $O(h)$. Consequently Δ_n is at least $O(\sqrt{h})$, so that (10.44) guarantees the convergence of the method. It is, however, somewhat pessimistic as it does not demonstrate the observed $O(h^{3/2})$ convergence. A deeper analysis is necessary to show this, to treat general kernels, or to justify extrapolation. Our understanding of these matters is still quite incomplete.

To proceed beyond this point to study various more accurate methods is complicated and involves a myriad of technical details. Many authors have obtained limited special results, but no complete theory has yet been established.

10.5. Solving Abel equations in the presence of experimental errors. As was remarked before, equations of the Abel type occur most commonly in connection with the analysis of certain experimental measurements. Consequently we must consider the effect of fairly large perturbations on the right-hand side. Again, to get some simple insight, take the simple case of $\mu = \frac{1}{2}$, $k(t, s) = 1$. If in (10.8) we perturb $g(t_n)$ by an amount η, then the change in $F_{n-1/2}$ will be of order $\eta/\sqrt{h}$. It is a fairly straightforward task, using arguments similar to those used in § 10.4, to show that this is a general result. A perturbation of order η causes a change in the computed solution of order $\eta/\sqrt{h}$. While the situation here is not as bad as for equations with smooth kernels (where the dependence is of order η/h), it is still sufficiently severe to warrant careful attention when η is significant.

The question of the solution of the simple Abel equation has been studied extensively. The conclusion is that in applications the form (10.3) is preferable to (10.4). A satisfactory procedure is first to smooth $g(t)$ by some methods such as least squares or Fourier approximation. Product integration is then used to evaluate the integral, and finally the derivative is computed by using a stable numerical differentiation method. The question of how to approximate the derivative of a function in the presence of uncertainties has received some attention and reasonably good methods are available. For references see the notes at the end of this chapter.

Another observation is sometimes helpful. It may not be $f(t)$ which is of interest, but some physically meaningful quantities, expressed as linear functionals of f. In the second example in § 2.4, $f(r)$ represents the probability density for the radius of the spherical particles. One value of obvious practical interest is the expected volume of the particles. If we assume that $f(r) = 0$ for $r > 1$, and $\rho = 1$, then the expected volume is

$$\bar{V} = \frac{4\pi}{3} \int_0^1 t^3 f(t) \, dt. \qquad (10.45)$$

Solving (2.55) for $f(r)$ using (5.32) and substituting into (10.45) gives

$$\bar{V} = -\frac{4}{3} \int_0^1 t^3 \frac{d}{dt} \int_t^1 \frac{G(s)}{(s^2 - t^2)^{1/2}} \, ds \, dt. \tag{10.46}$$

Integration by parts then yields

$$\bar{V} = 4 \int_0^1 t^2 \int_t^1 \frac{G(s)}{(s^2 - t^2)^{1/2}} \, ds \, dt. \tag{10.47}$$

To simplify further, interchange the order of integration and carry out the t-integration explicitly. The final expression is

$$\bar{V} = \pi \int_0^1 s^2 G(s) \, ds. \tag{10.48}$$

We find that we need only a simple integration to get $\bar{V}$; if this is all that is needed, it is completely unnecessary to solve the integral equation. Also, we note that the computation of $\bar{V}$ through (10.48) is completely well-posed. In fact, if the errors in $g(t)$ are random and without bias, their effect will tend to cancel out altogether.

Notes on Chapter 10. Solution methods for the simple Abel equation, using equation (10.4) with various ways of approximating $g'(t)$ are given in [102], [154], [191], [193], [206], [222]. For the general Abel equation, the midpoint method is studied in Weiss and Anderssen [236], while Atkinson [13], Eggermont [90] and Weiss [237] contain results on the trapezoidal methods. Other questions, including higher order methods, are discussed in [29], [32], [39], [44], [46], [47], [54], [133], [137], [138].

The error analysis used in § 10.4 was suggested by Eggermont in [89].

For Abel equations in the presence of errors, Baev and Glasko [15] suggest regularization, while Balasubramanian et al. [23] use a least-squares method. A thorough study of the simple Abel equation, including the use of stable differentiation methods and the computation of functionals, was carried out by Anderssen and co-workers. Some of their results are summarized in [4], [7], [9], [10], [142], [143].

Stable numerical differentiation schemes are considered in [5] and [6].

Chapter 11

Integrodifferential Equations

A variety of integrodifferential equations occur in practical applications. In this chapter we consider in detail only one case, the equation

$$f'(t) = H\left(t, f(t), \int_0^t K(t, s, f(s))\, ds\right), \qquad (11.1)$$

with initial condition

$$f(0) = f_0. \qquad (11.2)$$

The classical Volterra population equation discussed in Chapter 2 is of this form.

As indicated in Example 4.4, if we can assume that the simple Lipschitz conditions

$$|K(t, s, u_1) - K(t, s, u_2)| \leq L_1 |u_1 - u_2|, \qquad (11.3)$$

$$|H(t, v_1, w) - H(t, v_2, w)| \leq L_2 |v_1 - v_2|, \qquad (11.4)$$

$$|H(t, v, w_1) - H(t, v, w_2)| \leq L_3 |w_1 - w_2|, \qquad (11.5)$$

hold for $0 \leq s \leq t \leq T$ and all u, v, w, then (11.1), subject to the condition (11.2), has a unique continuously differentiable solution.

Equation (11.1) can be considered a generalization of the ordinary differential equation

$$f'(t) = H(t, f(t)). \qquad (11.6)$$

Consequently the numerical methods for (11.1) bear a great deal of similarity to the results for initial value problems in ordinary differential equations.

In §§ 11.1 to 11.4 we outline some of the numerical methods which have been studied. A quite simple method is analyzed in detail in § 11.1. In § 11.2 we discuss the class of linear multistep methods and show how the classical *Dahlquist* theory carries over to integrodifferential equations. Block-by-block methods are also suitable for this case and are discussed in § 11.3.

Section 11.4 considers the question of numerical stability. The chapter concludes with a very brief discussion of some important types of equations not of the form (11.1).

11.1. A simple numerical method. For the development of numerical methods it is convenient to rewrite (11.1) as

$$f'(t) = H(t, f(t), z(t)), \qquad (11.7)$$

$$z(t) = \int_0^t K(t, s, f(s))\, ds. \qquad (11.8)$$

We then integrate (11.7) from t_{n-1} to t_n to give

$$f(t_n) = f(t_{n-1}) + \int_{t_{n-1}}^{t_n} H(s, f(s), z(s))\, ds. \qquad (11.9)$$

Replacing this integral and the integral in the definition of $z(t)$ by a numerical integration rule yields a formula for computing the approximate solution F_n. For example, if we use the trapezoidal rule for both integrals, we get

$$F_n = F_{n-1} + \frac{h}{2}\{H(t_{n-1}, F_{n-1}, Z_{n-1}) + H(t_n, F_n, Z_n)\}, \qquad (11.10)$$

$$F_0 = f_0,$$

where

$$Z_0 = 0,$$

$$Z_n = \frac{h}{2} k(t_n, t_0, F_0) + h \sum_{i=1}^{n-1} k(t_n, t_i, F_i) + \frac{h}{2} k(t_n, t_n, F_n),$$

$$n = 1, 2, \ldots. \qquad (11.11)$$

Equations (11.10) and (11.11) can be solved successively for $F_1, F_2, \ldots$. In the nonlinear case F_n is defined implicitly, but as usual, a unique solution exists for sufficiently small h.

Example 11.1. The equation

$$f'(t) = 1 + f(t) - te^{-t^2} - 2\int_0^t tse^{-f^2(s)}\, ds, \qquad (11.12)$$

whose exact solution is $f(t) = t$, was solved using (11.10) and (11.11) with various values of h. At each step, the nonlinear equation was solved by successive substitution, starting with

$$F_n^{(0)} = F_{n-1},$$

TABLE 11.1
Solution of (11.12) by the trapezoidal method.

t	$h=0.2$	$h=0.1$	$h=0.05$
0.2	0.20001731	0.20000259	0.20000054
0.4	0.40016431	0.40003414	0.40000811
0.6	0.60069227	0.60015777	0.60003851
0.8	0.80193268	0.80045560	0.80011223
1.0	1.00424592	1.00101469	1.00025084

and iterating until
$$|F_n^{(j)} - F_n^{(j-1)}| \leq 10^{-8}.$$

The last value of $F_n^{(j)}$ was then taken as a sufficiently close approximation to F_n.

The results are shown in Table 11.1. The apparent order of convergence is two, which is not surprising because of the use of the trapezoidal method.

For the error analysis of such methods, we need another basic result on sequences generated by sums.

THEOREM 11.1. *Suppose that the sequence $\{\xi_n\}$ satisfies*

$$|\xi_n| \leq A |\xi_{n-1}| + B \sum_{i=0}^{n-1} |\xi_i| + C, \tag{11.13}$$

with $A > 1$, positive B, C, and $\xi_0 = 0$. Then

$$|\xi_n| \leq \frac{(A+nB)^n - 1}{A-1} C. \tag{11.14}$$

Proof. Assume that (11.14) is satisfied for $i = 1, 2, \ldots, n-1$. Then

$$|\xi_n| \leq A \frac{[A+(n-1)B]^{n-1} - 1}{A-1} C + B \sum_{i=0}^{n-1} \frac{(A+iB)^i - 1}{A-1} C + C$$

$$\leq A \frac{(A+nB)^{n-1} - 1}{A-1} C + nB \frac{(A+nB)^{n-1} - 1}{A-1} C + C$$

$$= \frac{(A+nB)^n - nB - 1}{A-1} C \leq \frac{(A+nB)^n - 1}{A-1} C.$$

For $n = 1$ we have $|\xi_1| \leq C$, so that (11.14) is satisfied. Hence by induction the inequality holds for all n. □

With this result, the error analysis for (11.10) is straightforward. We sketch it briefly, omitting some details.

THEOREM 11.2. *If conditions (11.2)–(11.4) are satisfied, and if in addition H and K are twice continuously differentiable with respect to all arguments,* then the approximate solution defined by (11.10) and (11.11) converges to the true solution of (11.1) with order two.

Proof. We set as usual $\varepsilon_n = F_n - f(t_n)$, and use the further shorthand notation

$$H_n = H(t_n, F_n, Z_n) = H\left(t_n, F_n, h \sum_{i=0}^{n}{}'' K(t_n, t_i, F_i)\right), \quad (11.15)$$

$$\tilde{H}_n = H\left(t_n, f(t_n), h \sum_{i=0}^{n}{}'' K(t_n, t_i, f(t_i))\right), \quad (11.16)$$

where the double primes on the summation sign indicate that the first and last terms are to be halved. Then, subtracting (11.9) from (11.10), we have

$$\varepsilon_n = \varepsilon_{n-1} + \frac{h}{2}\{H_{n-1} + H_n\} - \int_{t_{n-1}}^{t_n} H(s, f(s), z(s))\, ds$$

$$= \varepsilon_{n-1} + \frac{h}{2}\{H_{n-1} - \tilde{H}_{n-1}\} + \frac{h}{2}\{H_n - \tilde{H}_n\} + \delta_n, \quad (11.17)$$

where δ_n is the integration error

$$\delta_n = \frac{h}{2}\{\tilde{H}_{n-1} + \tilde{H}_n\} - \int_{t_{n-1}}^{t_n} H(t, f(t), z(t))\, dt. \quad (11.18)$$

Now, from (11.15) and (11.16)

$$H_n - \tilde{H}_n = H\left(t_n, F_n, h \sum_{i=0}^{n}{}'' K(t_n, t_i, F_i)\right) - H\left(t_n, f(t_n), h \sum_{i=0}^{n}{}'' K(t_n, t_i, F_i)\right)$$

$$+ H\left(t_n, f(t_n), h \sum_{i=0}^{n}{}'' K(t_n, t_i, F_i)\right)$$

$$- H\left(t_n, f(t_n), h \sum_{i=0}^{n}{}'' K(t_n, t_i, f(t_i))\right),$$

so that, applying the Lipschitz conditions (11.3)–(11.5), we have

$$|H_n - \tilde{H}_n| \leq L_2 |\varepsilon_n| + h L_1 L_3 \sum_{i=0}^{n}{}'' |\varepsilon_i|.$$

Substituting this inequality into (11.17), we get

$$|\varepsilon_n| \leq |\varepsilon_{n-1}| + \frac{h}{2}\left\{L_2 |\varepsilon_{n-1}| + h L_1 L_3 \sum_{i=0}^{n-1}{}'' |\varepsilon_i|\right\}$$

$$+ \frac{h}{2}\left\{L_2 |\varepsilon_n| + h L_1 L_3 \sum_{i=0}^{n}{}'' |\varepsilon_i|\right\} + |\delta_n|.$$

For sufficiently small h this implies that

$$|\varepsilon_n| \leq A |\varepsilon_{n-1}| + h^2 B \sum_{i=0}^{n-1} |\varepsilon_i| + \Delta_n$$

with

$$A = \frac{1+(h/2)L_2}{1-(h/2)L_2-(h^2/2)L_1L_3},$$

$$B = \frac{L_1L_3}{1-(h/2)L_2-(h^2/2)L_1L_3},$$

$$\Delta_n = \frac{|\delta_n|}{1-(h/2)L_2-(h^2/2)L_1L_3}.$$

Applying Theorem 11.1 it follows that

$$|\varepsilon_n| \leq \frac{(A+nh^2B)^n - 1}{A-1} \max_{1 \leq i \leq n} |\Delta_i|. \tag{11.19}$$

It is now easily shown that

$$A - 1 = O(h), \quad (A + nh^2B)^n = O(1),$$

so that

$$|\varepsilon_n| = O\left(\frac{|\delta_n|}{h}\right). \tag{11.20}$$

For sufficiently smooth H and K the use of the trapezoidal rule implies that

$$\delta_n = O(h^3) \tag{11.21}$$

which completes the proof. □

Higher order methods can be constructed along similar lines. For example, if we integrate (11.1) from t_{n-2} to t_n and replace the result by Simpson's rule we obtain

$$F_n = F_{n-2} + \frac{h}{3}\{H(t_{n-2}, F_{n-2}, Z_{n-2}) + 4H(t_{n-1}, F_{n-1}, Z_{n-1}) + H(t_n, F_n, Z_n)\}, \tag{11.22}$$

with

$$Z_n = h \sum_{i=0}^{n} w_{ni} K(t_n, t_i, F_i). \tag{11.23}$$

The weights w_{ni} are determined by whatever numerical integration rule we choose. Normally one would choose it to have the same order of accuracy as Simpson's rule.

We will not further pursue the question of convergence of this method since it is a special case of a general theory which we describe next.

11.2. Linear multistep methods.

When H is independent of z, then (11.1) reduces to an ordinary differential equation. In this instance the approximate methods described so far reduce to well-known methods for solving ordinary differential equations. Thus (11.10) becomes

$$F_n = F_{n-1} + \frac{h}{2}\{H(t_{n-1}, F_{n-1}) + H(t_n, F_n)\}, \tag{11.24}$$

while (11.22) reduces to

$$F_n = F_{n-2} + \frac{h}{3}\{H(t_{n-2}, F_{n-2}) + 4H(t_{n-1}, F_{n-1}) + H(t_n, F_n)\}. \tag{11.25}$$

Both of these are simple examples of the so-called *multistep methods*; (11.24) is of the *Adams–Moulton* type while (11.25) is an instance of the *Milne–Simpson* method. A great deal of work has been done in connection with multistep methods, with the Dahlquist stability theory as one of the major results. As we now indicate, it is possible to extend much of this theory to integrodifferential equations.

As a general form for the multistep methods for integrodifferential equations we will take

$$\alpha_k F_{n+k} + \alpha_{k-1} F_{n+k-1} + \cdots + \alpha_0 F_n$$
$$= h\{\beta_k H_{n+k} + \beta_{k-1} H_{n+k-1} + \cdots + \beta_0 H_n\}, \tag{11.26}$$

where

$$H_n = H(t_n, F_n, Z_n), \tag{11.27}$$

$$Z_n = h \sum_{i=0}^{n} w_{ni} K(t_n, t_i, F_i). \tag{11.28}$$

These equations will generally be valid only for $n = r, r+1, \ldots,$ and to completely determine the solution we then need $r+k$ starting values. If these are available, then (11.26)–(11.28) can be solved for F_{n+k} at each step (provided that h is sufficiently small).

Because the theory we are about to describe closely parallels the corresponding theory for ordinary differential equations we use terminology and notation customary in this field. We first introduce two polynomials $\rho(x)$ and $\sigma(x)$ defined by

$$\rho(x) = \alpha_k x^k + \alpha_{k-1} x^{k-1} + \cdots + \alpha_0, \tag{11.29}$$

$$\sigma(x) = \beta_k x^k + \beta_{k-1} x^{k-1} + \cdots + \beta_0, \tag{11.30}$$

then use the following definitions.

DEFINITION 11.1. A multistep method of the form (11.26)–(11.28) satisfies the *consistency conditions* if

(i) $$\rho(1) = 0, \qquad (11.31)$$

(ii) $$\rho'(1) = \sigma(1), \qquad (11.32)$$

(iii) the weights w_{ni} are uniformly bounded, that is,

$$\sup_h \max_{i \le n} |w_{ni}| \le W < \infty,$$

and are such that

$$\lim_h \left(\int_0^x f(t)\, dt - h \sum_{i=0}^n w_{ni} f(t_i) \right) = 0, \qquad (11.33)$$

for every continuous function $f(t)$.

DEFINITION 11.2. A multistep method satisfies the *stability condition* if all of the roots of the polynomial $\rho(x)$ lie in or on the circle $|x| = 1$ in the complex plane, and roots of modulus one are simple.

The reader can easily verify that the two methods above satisfy the consistency and stability conditions. Using these definitions we can then extend the classical Dahlquist theorem for ordinary differential equations to integrodifferential equations.

THEOREM 11.3. *If a multistep method of the form* (11.26) *satisfies the consistency condition of Definition* 11.1 *and the stability condition of Definition* 11.2, *and if the starting errors tend to zero as* $h \to 0$, *then the method is convergent.*

Proof. The arguments follow closely those for differential equations found in most books on the numerical solution of differential equations. For integrodifferential equations there are a few additional details which will not be reproduced here. Explicit arguments can be found in several references on this topic, for example in [166]. □

The concept of consistency, introduced in Chapter 7, is closely related to the consistency condition of Definition 11.1. In spite of the different appearance, the two concepts are based on the same idea. From Definition 7.3 we see that consistency is defined by subtracting the original equation from that obtained by using a numerical integration rule in place of the integral. If we apply this idea to the integrodifferential equation, we are led to defining the consistency error as

$$\begin{aligned}\delta(h, t_n) = &\frac{1}{h} \{\alpha_k f(t_{n+k}) + \alpha_{k-1} f(t_{n+k-1}) + \cdots + \alpha_0 f(t_n)\} \\ &+ \{\beta_k \tilde{H}_{n+k} + \beta_{k-1} \tilde{H}_{n+k-1} + \cdots + \beta_0 \tilde{H}_n\} \\ &- \left\{ f'(t) - H\left(t, f(t), \int_0^t K(t, s, f(s))\, ds\right) \right\},\end{aligned} \qquad (11.34)$$

for some t in $[t_n, t_{n+k}]$. We then want to show that consistency in the sense of Definition 11.1 implies that

$$\lim_h \delta(h, t_n) = 0.$$

The terms in the last curly bracket in (11.34) are identically zero for all t, so that the equation can be rewritten as

$$\delta(h, t_n) = \frac{1}{h} \{\alpha_k f(t_{n+k}) + \alpha_{k-1} f(t_{n+k-1}) + \cdots + \alpha_0 f(t_n)$$
$$- h\beta_k f'(t_{n+k}) - h\beta_{k-1} f'(t_{n+k-1}) - \cdots - h\beta_0 f'(t_n)\}$$
$$+ \left\{\beta_k H\left(t_{n+k}, f(t_{n+k}), \int_0^{t_{n+k}} K(t_{n+k}, s, f(s))\, ds\right) + \cdots \right.$$
$$\left. + \beta_0 H\left(t_n, f(t_n), \int_0^{t_n} K(t_n, s, f(s))\, ds\right) - \beta_k \tilde{H}_{n+k} - \cdots - \beta_0 \tilde{H}_n\right\}.$$
(11.35)

A Taylor expansion of the first group of terms in (11.35) shows that the first two consistency conditions (11.31) and (11.32) are sufficient to make these terms vanish as $h \to 0$. Condition (11.33) implies that the second group of terms will also tend to zero in the limit. Thus, we have that if (11.31)–(11.33) are satisfied then

$$\lim_h \delta(h, t_n) = 0.$$

The order of consistency of the multistep (11.26) can be defined via the consistency error $\delta(h, t_n)$ in (11.34). If

$$\delta(h, t_n) = O(h^p),$$

we say that the method is consistent of order p. As expected, the order of consistency and the order of convergence are essentially the same.

THEOREM 11.4. *Assume that the multistep method* (11.26) *satisfies the stability condition of Definition* 11.2 *and is consistent. Let* $\eta(h)$ *be the absolute sum of the starting errors, that is,*

$$\eta(h) = \sum_{i=0}^{r+k-1} |F_i - f(t_i)|. \tag{11.36}$$

Then there exist constants c_1 *and* c_2, *independent of* n *and* h, *such that*

$$|F_n - f(t_n)| \leq c_1 |\delta(h, t_n)| + c_2 \eta(h). \tag{11.37}$$

Proof. Again the arguments are similar to the differential equations case, with some additional complications. For details, consult the reference cited in Theorem 11.3. □

However, even though the consistency condition introduced here is equivalent to the one encountered before, the same cannot be said for the stability condition. The stability condition of Definition 11.2 has little to do with the question of numerical stability discussed in § 7.4. A method which does not satisfy the stability condition of Definition 11.2 is nonconvergent, so the question of numerical stability is irrelevant. On the other hand, a method which does satisfy the stability condition, although convergent, may still be numerically unstable. We will take a brief look at the numerical stability problem for integrodifferential equations in § 11.4.

The results developed here make it easy to construct multistep methods for equations of the form (11.1). We simply combine a multistep method for ordinary differential equations, such as an Adams–Moulton formula, with an appropriate numerical integration rule. This works because the consistency conditions (11.31) and (11.32) and the stability condition refer only to the discretization of the derivative and are identical with the conditions imposed on multistep methods for ordinary differential equations.

The consistency condition (11.33) assures that the numerical integration used to replace the integral is convergent.

Example 11.2. We can choose

$$\alpha_2 = 1, \quad \beta_2 = \tfrac{5}{12}, \quad w_{n0} = w_{nn} = \tfrac{5}{12} \ (n \geq 3),$$
$$\alpha_1 = -1, \quad \beta_1 = \tfrac{2}{3}, \quad w_{n1} = w_{n,n-1} = \tfrac{13}{12},$$
$$\alpha_0 = 0, \quad \beta_0 = -\tfrac{1}{12}, \quad w_{ni} = 1 \quad \text{otherwise.}$$

This method is a combination of the third order Adams–Moulton method with the third order Gregory integration formula. It requires five starting values $F_0, \ldots, F_4$. Its order of convergence is three.

11.3. Block-by-block methods. The block-by-block methods for integral equations can be modified to work for integrodifferential equations. This yields a class of methods which can be used either to provide starting values for the multistep methods or simply as techniques for solving the whole problem.

To construct block-by-block methods we start from the integrated form of (11.1). In particular, if for some $p > 1$ we integrate (11.1) from t_n to t_{n+p}, then

$$f(t_{n+p}) = f(t_n) + \int_{t_n}^{t_{n+p}} H(t, f(s), z(s)) \, ds, \tag{11.38}$$

$$z(t) = \int_0^t k(t, s, f(s)) \, ds. \tag{11.39}$$

Block-by-block methods are then constructed by appropriate combinations of numerical integration and interpolation. We again present the method based on Simpson's rule and quadratic interpolation as an example. Using $p=1$ and $p=2$ in (11.38) and replacing the integral by an appropriate Simpson's rule we have

$$F_{2n+1} = F_{2n} + \frac{h}{6}\{H(t_{2n}, F_{2n}, Z_{2n}) + 4H(t_{2n+1/2}, F_{2n+1/2}, Z_{2n+1/2})$$
$$+ H(t_{2n+1}, F_{2n+1}, Z_{2n+1})\}, \quad (11.40)$$

$$F_{2n+2} = F_{2n} + \frac{h}{3}\{H(t_{2n}, F_{2n}, Z_{2n}) + 4H(t_{2n+1}, F_{2n+1}, Z_{2n+1})$$
$$+ H(t_{2n+2}, F_{2n+2}, Z_{2n+2})\}. \quad (11.41)$$

The Z_i are then computed by applying Simpson's rule to (11.39),

$$Z_{2n+1} = \frac{h}{3} \sum_{i=0}^{2n} w_i K(t_{2n+1}, t_i, F_i)$$
$$+ \frac{h}{6}\{K(t_{2n+1}, t_{2n}, F_{2n}) + 4K(t_{2n+1}, t_{2n+1/2}, F_{2n+1/2})$$
$$+ K(t_{2n+1}, t_{2n+1}, F_{2n+1})\},$$

$$Z_{2n+2} = \frac{h}{3} \sum_{i=0}^{2n+2} w_i K(t_{2n+2}, t_i, F_i). \quad (11.42)$$

Here $t_{n+1/2} = t_n + h/2$ and $\{w_i\}$ is the set of Simpson's rule weights $\{1, 4, 2, 4, \ldots, 4, 2, 4, 1\}$. The extraneous values $F_{2n+1/2}$ and $Z_{2n+1/2}$ are approximated by quadratic interpolation as

$$F_{2n+1/2} = \tfrac{3}{8} F_{2n} + \tfrac{3}{4} F_{2n+1} - \tfrac{1}{8} F_{2n+2}, \quad (11.43)$$

$$Z_{2n+1/2} = \tfrac{3}{8} Z_{2n} + \tfrac{3}{4} Z_{2n+1} - \tfrac{1}{8} Z_{2n+2}. \quad (11.44)$$

Equations (11.40)–(11.44) constitute an implicit set of equations for F_{2n+1} and F_{2n+2}. Starting with $F_0 = f_0$ we can solve these for $F_1, F_2, F_3, F_4, \ldots$, obtaining a block of two values at each step. This method has fourth order accuracy, as is easily proved. It is also easy to see that one can generalize the above idea to construct methods of arbitrarily high order.

11.4. Numerical stability. To investigate the numerical stability of a method we have two alternatives: we can carry out an asymptotic analysis to determine the behavior of the dominant error term, or we can apply the method to a simple, completely solvable case. The latter approach, being somewhat simpler, seems to be preferred by writers on integrodifferential equations and we will follow this tradition. A suitable equation for the analysis is

$$f'(t) = -(\lambda + \mu)f(t) - \lambda\mu \int_0^t f(s)\,ds. \quad (11.45)$$

The solution of this equation is easily seen to be of the form

$$f(t) = c_1 e^{-\lambda t} + c_2 e^{-\mu t}, \tag{11.46}$$

so that if $\lambda > 0$, $\mu > 0$, $f(t) \to 0$ as $t \to \infty$. A numerical method will then be called stable if, when applied to (11.45) with positive λ and μ, and fixed h, it yields an approximate solution F_n for which

$$\lim_{n \to \infty} F_n = 0. \tag{11.47}$$

The values of h for which (11.47) holds are the stability region.

Let us apply this criterion to the trapezoidal method (11.10) and (11.11). For equation (11.46), the approximating scheme gives

$$F_n = F_{n-1} - \frac{h(\lambda + \mu)}{2}\{F_{n-1} + F_n\} - \frac{h^2 \lambda \mu}{2}\left\{\sum_{i=0}^{n-1}{}'' F_i + \sum_{i=0}^{n}{}'' F_i\right\}, \tag{11.48}$$

where the double prime denotes that the first and last terms in the sum are halved. The solution to this equation becomes more obvious if we subtract from it the same equation with n replaced by $n-1$. Doing this, cancelling terms, and rearranging we get

$$\left(1 + \frac{h(\lambda + \mu)}{2} + \frac{h^2 \lambda \mu}{2}\right) F_n - \left(2 - \frac{h^2 \lambda \mu}{2}\right) F_{n-1} + \left(1 - \frac{h(\lambda + \mu)}{2} + \frac{h^2 \lambda \mu}{4}\right) F_{n-2} = 0. \tag{11.49}$$

Therefore the solution to (11.48) is

$$F_n = c_1 \rho_1^n + c_2 \rho_2^n,$$

where ρ_1 and ρ_2 are the roots of the characteristic polynomial

$$\left(1 + \frac{h(\lambda + \mu)}{2} + \frac{h^2 \lambda \mu}{4}\right) \rho^2 - \left(2 - \frac{h^2 \lambda \mu}{2}\right) \rho + \left(1 - \frac{h(\lambda + \mu)}{2} + \frac{h^2 \lambda \mu}{4}\right) = 0.$$

Solving this quadratic equation gives roots ρ_1 and ρ_2 as

$$\rho_1 = \frac{1 - h(\lambda - \mu)/2 - h^2 \lambda \mu/4}{1 + h(\lambda + \mu)/2 + h^2 \lambda \mu/4},$$

$$\rho_2 = \frac{1 + h(\lambda - \mu)/2 - h^2 \lambda \mu/4}{1 + h(\lambda + \mu)/2 + h^2 \lambda \mu/4}.$$

Since both roots have magnitude less than one for any positive h, λ, μ we see that

$$\lim_{n \to \infty} F_n = 0$$

for every $h > 0$, therefore, the method is A-stable.

A similar process can be carried out for the other multistep methods and regions of stability can be computed.

One easy way of obtaining a characterization of the stability of multistep methods is to note that if a method is to be considered stable it ought certainly to be so in the special case when (11.1) reduces to a differential equation. In that case, the corresponding multistep method reduces to a method for ordinary differential equations, and much is known about the stability in that case. For example, if we consider the method defined by (11.22) and (11.23) then when H is independent of Z, the equations reduce to the well-known Milne–Simpson method (11.25). It is known that this method is numerically unstable. Therefore, we can characterize method (11.22), (11.23) as numerically unstable without having to carry out a lengthy analysis.

11.5. Other types of integrodifferential and functional equations. In principle it is possible to extend the techniques discussed to other types of integrodifferential equations. The discretization and design of algorithms is quite elementary, but the analysis tends to get involved. Actually, there is little reason for pursuing this line of thought. The general equivalence between integrodifferential equations and systems of Volterra equations already discussed in earlier chapters provides an immediate way for the construction of numerical algorithms. The reduction of integrodifferential equations of arbitrarily high order to systems of Volterra equations is an elementary matter. Consider, for example, the general nonlinear case

$$f^{(m)}(t) = H\Big(t, f(t), f'(t), \ldots, f^{(m-1)}(t), \int_0^t K(t, s, f(s), \ldots, f^{(m-1)}(s))\, ds\Big),$$
(11.50)

with given initial conditions

$$f(0) = \alpha_0,$$
$$f'(0) = \alpha_1,$$
$$\vdots$$
$$f^{(m-1)}(0) = \alpha_{m-1}.$$

Integrating (11.50) once, we get

$$f^{(m-1)}(t) = \alpha_{m-1} + \int_0^t H(s, f(s), \ldots, f^{(m-1)}(s), z_m(s))\, ds, \quad (11.51)$$

where

$$z_m(t) = \int_0^t K(t, s, f(s), \ldots, f^{(m-1)}(s))\, ds. \quad (11.52)$$

If we now introduce variables $z_i = f^{(i)}$, $i = 0, 1, \ldots, m-1$, we arrive at the system

$$z_m(t) = \int_0^t K(t, s, z_0(s), \ldots, z_{m-1}(s)) \, ds, \qquad (11.53)$$

$$z_{m-1}(t) = \alpha_{m-1} + \int_0^t H(s, z_0(s), \ldots, z_{m-1}(s), z_m(s)) \, ds, \qquad (11.54)$$

$$z_i(t) = \alpha_i + \int_0^t z_{i+1}(s) \, ds, \qquad i = 0, 1, \ldots, m-2. \qquad (11.55)$$

Numerically this can be solved with any of the techniques described in Chapter 7. It is of course possible to start directly from (11.50) and develop special methods analogous to those for differential equations of higher order. However, the analysis for this appears to be difficult, with little to be gained by it.

In many applications there arise equations which are somewhat different from (11.1). Equation (11.1) and a large number of other equations can be considered special instances of the general Volterra *functional differential equation*

$$f'(t) = H(t, f_t), \qquad (11.56)$$

where f_t maps the function f defined on $-\infty < s \leq t$ into the space of real numbers. Obviously, (11.1) is a special case of (11.56). So is the simple *delay differential equation*

$$f'(t) = H(t, f(t-\alpha)), \qquad (11.57)$$

where α is some fixed positive number. In extending our techniques to (11.56) some complications may be encountered. For instance, if we want to solve (11.57) for $t \geq 0$, we need to know $f(t)$ for $-\alpha \leq t < 0$. This points out that the set of necessary initial conditions can be more complicated than in (11.1). Furthermore, again referring to (11.57), unless $f'(0-) = H(0, f(-\alpha))$, the solution has a discontinuous derivative at $t = 0$. This, in turn, generates a discontinuous second derivative at $s = \alpha$, and so on. Since numerical methods are generally designed on the assumption that the solution is very smooth, this may cause trouble. Nevertheless, with a certain amount of care, it is possible to extend known techniques for Volterra integrodifferential equations to general functional differential equations. The resulting numerical methods are much like the methods we have presented here.

Notes on Chapter 11. A discussion of the theory of multistep methods, including proof of Theorems 11.3 and 11.4 is given in Linz [166]. Other related questions are studied by Mocarsky [192] and McKee [182]. Piecewise polynomial and spline approximations are described in [45] and [122], while [100] and [210] give some other methods that require no starting values. Equations with singular kernels are the subject of study in [177], [184], [201], [213]. Chang and Day [64] study an equation somewhat different from that considered here.

The question of numerical stability for integrodifferential equations is treated in Brunner and Lambert [48] and Matthys [180].

There is an extensive literature on the numerical solution of functional differential equations. Some representative papers are [72], [65], [124], [144], [148], [149], [227], [240].

Partial integrodifferential equations of Volterra type are also occasionally encountered in practice. Ford [103] and Neta [194] discuss some problems and their numerical solution, but this has not been systematically studied and is presently not well understood.

Chapter 12

Some Computer Programs

While there exist many methods for the solution of Volterra equations, there are very few published computer programs. This is not uncommon for numerical analysis in general; theoretical or descriptive papers greatly outnumber those giving actual programs. Apart from the fact that programming is on the whole not considered as creative as theorem proving, there may be other reasons for most authors' reluctance to include programs in their publications. One is that the methods are constructed for the simple, prototype cases. Applications often involve equations somewhat different from these, so that the programs can rarely be used without modification. The designer consequently has to choose between sophisticated and highly efficient programs for standard equations, and flexible, easily modifiable, but inefficient algorithms. Another unrewarding decision which has to be made is what language to use. There are advantages and disadvantages to every possible choice. Nevertheless, even though one cannot hope to please everyone with one's decisions, it is reasonable to provide at least something which may occasionally be helpful. It is the spirit of these observations which guided the decisions for the programs in this chapter.

Our major decision was to use essentially the simplest nontrivial algorithms, namely, the trapezoidal method for equations of the second kind, and the midpoint method for equations of the first kind. One reason for this selection is that the methods are well understood and have good stability properties. They should deliver acceptable results in a variety of settings. Another advantage is that their simplicity makes modifications reasonably easy. Their main disadvantage is their poor accuracy when compared to higher order methods. This may not be too serious. Highly accurate results are rarely needed; if they are, extrapolation techniques can sometimes be used to improve the results with a minimal effort.

To make the programs useful in many circumstances, the implementations were done for systems of equations, rather than single equations. For equations of the second kind the nonlinear problem is treated. However, because of theoretical difficulties, only linear equations of the first kind are considered. To cover most of the standard types discussed in this book, four programs are provided. Two are for nonlinear systems of the second kind, using the regular trapezoidal and the product trapezoidal methods. For equations of the first kind, the programs are based on the midpoint and the product midpoint methods. The last of these can be used for Abel-type equations. Since all programs are for systems, they can be used to solve integrodifferential equations as well.

The most debatable decision was the selection of the programming language. FORTRAN, because of its wide usage, would probably have been a popular choice. Unfortunately, it is difficult to program cleanly in this language. Since program modifications are to be expected, a language with a better structure is preferable. Pascal was chosen as a compromise between ease of programming and availability. An attempt was made to keep the programs as simple as possible, so that an experienced programmer should have little difficulty in translating them to another language.

The programs provided here are aimed at the casual user. Their design emphasizes simplicity and generality over computational efficiency. For the repeated computation of accurate solutions of large systems they are relatively inefficient.

12.1. The trapezoidal method for systems of the second kind. We consider here a system of m equations of the form

$$\mathbf{f}(t) = \mathbf{g}(t) + \int_0^t \mathbf{K}(t, s, \mathbf{f}(s))\, ds, \qquad (12.1)$$

where $\mathbf{f}$, $\mathbf{g}$ and $\mathbf{K}$ are vectors with m components, as given in equations (1.9) and (1.10). The trapezoidal approximation method (7.3), when written for such a system, becomes

$$\mathbf{F}_n = \mathbf{g}(t_n) + h\left\{\tfrac{1}{2}\mathbf{K}(t_n, t_0, \mathbf{F}_0) + \sum_{j=1}^{n-1} \mathbf{K}(t_n, t_j, \mathbf{F}_j) + \tfrac{1}{2}\mathbf{K}(t_n, t_n, \mathbf{F}_n)\right\},$$

$$(12.2)$$

where $\mathbf{F}_n$ is an m-component vector

$$\mathbf{F}_n = \begin{pmatrix} (F_n)_1 \\ (F_n)_2 \\ \vdots \\ (F_n)_m \end{pmatrix}. \qquad (12.3)$$

In this notation, $(F_n)_i$ is the approximation to $f_i(t_n)$.

With $\mathbf{F}_0 = \mathbf{g}(t_0)$, equation (12.2) is a nonlinear system to be solved in a stepwise fashion for $\mathbf{F}_1, \mathbf{F}_2, \ldots$. Such a system can be solved in various ways. Here we will use Newton's method.

If we write (12.2) as

$$\mathbf{F}_n - \frac{h}{2}\mathbf{K}(t_n, t_n, \mathbf{F}_n) - \boldsymbol{\chi}_n = 0, \qquad (12.4)$$

where

$$\boldsymbol{\chi}_n = \mathbf{g}(t_n) + h\left\{\tfrac{1}{2}\mathbf{K}(t_n, t_0, \mathbf{F}_0) + \sum_{j=1}^{n-1} \mathbf{K}(t_n, t_j, \mathbf{F}_j)\right\}, \qquad (12.5)$$

then the Newton iterates are

$$\mathbf{F}_n^{(i+1)} = \mathbf{F}_n^{(i)} - J^{-1}(\mathbf{F}_n^{(i)})\left\{\mathbf{F}_n^{(i)} - \frac{h}{2}\mathbf{K}(t_n, t_n, \mathbf{F}_n^{(i)}) - \boldsymbol{\chi}_n\right\}. \qquad (12.6)$$

Here $\mathbf{F}_n^{(i)}$ stands for the value of $\mathbf{F}_n$ at the ith iteration, and J is the Jacobian matrix with elements

$$[J(\mathbf{u})]_{ij} = \delta_{ij} - \frac{h}{2}\frac{\partial K_i(t_n, t_n, \mathbf{u})}{\partial u_j}. \qquad (12.7)$$

The iterations (12.6) will be carried out until

$$\|\mathbf{F}_n^{(i+1)} - \mathbf{F}_n^{(i)}\| \leq \varepsilon,$$

where ε is some assigned tolerance. The last iterate will be taken as $\mathbf{F}_n$. This value will be used as the first guess, $\mathbf{F}_{n+1}^{(0)}$, for the next step. This algorithm is implemented in the following Pascal procedure VOLT2.

The procedure VOLT2 is called by

$$\text{VOLT2}(h, N)$$

where

h(of type real) = stepsize,
N(of type integer) = total number of steps taken.

The calling program must contain the following declarations:

```
const m = (size of system);
type MATRIX = array [1..m, 1..m] of real;
     VECTOR = array [1..m] of real;
  var t: array [0..NX] of real;
      F: array [0..NX] of VECTOR
```

where NX is an integer constant equal to or larger than N. The results produced by VOLT2 will be stored in t and F, with t_0 in $t[0]$, t_1 in $t[1]$, and so on. The vector $F[n]$ represents $\mathbf{F}_n$, consequently $F[n][i]$ is the value of $(F_n)_i$ in (12.3).

The user must also provide three real functions g, K and KDIFF, so that (with i, j integer, t, s real, and u of type VECTOR),

$$g(i, t) = g_i(t),$$
$$K(i, t, s, u) = K_i(t, s, u[1], \ldots, u[m]),$$
$$KDIFF(i, j, t, s, u) = \frac{\partial K_i(t, s, u[1], \ldots, u[m])}{\partial u[j]}.$$

The procedure VOLT2 requires a procedure for the solution of a linear algebraic system. For completeness, we include a simple Gaussian elimination algorithm SOLVE in the code below.

Example 12.1. The system

$$f_1(t) = e^{-t} - \frac{t^2}{2} + \int_0^t e^{s-t} f_1^2(s) \, ds + \int_0^t f_2(s) \, ds, \qquad (12.8)$$

$$f_2(t) = t - \frac{t^4}{24} + \int_0^t (t-s) \frac{f_2^2(s)}{1 + f_1^2(s)} \, ds, \qquad (12.9)$$

has exact solution $f_1(t) = 1$ and $f_2(t) = t$. To solve this with stepsize $h = 0.05$ in the interval $[0, 0.5]$ define

$$\text{const } m = 2$$

and use

$$\text{VOLT2}(0.05, 10)$$

with the code for g, K, and KDIFF shown in Fig. 12.1. The listing for VOLT2 is given in Fig. 12.2.

```
function g(i:integer;t:real):real;
begin
  case i of
    1: g:=exp(-t)-t*t/2;
    2: g:=t-t*t*t*t/24
  end
end;

function K(i:integer;t,s:real;u:VECTOR):real;
begin
  case i of
    1: K:=exp(s-t)*u[1]*u[1]+u[2];
    2: K:=(t-s)*u[2]*u[2]/(1+u[1]*u[1])
  end
end;

function KDIFF(i,j:integer;t,s:real;u:VECTOR):real;
begin
  case i of
    1: case j of
         1: KDIFF:=2*exp(s-t)*u[1];
         2: KDIFF:=1
       end;
    2: case j of
         1: KDIFF:=-2*(t-s)*u[1]*u[2]*u[2]/sqr(1+u[1]*u[1]);
         2: KDIFF:=2*(t-s)*u[2]/(1+u[1]*u[1])
       end
  end
end;
```

FIG. 12.1. *Code for functions g and K.*

```
procedure SOLVE(var A:MATRIX; var B, X: VECTOR);
{ procedure for solving an mxm linear system.
          A*X =B
  The calling program must define the integer
  constant m and types
     MATRIX=array[1..m,1..m] of real;
     VECTOR=array[1..m] of real
  The algorithm is Gaussian elimination with
  simple pivoting. If any pivot is less than
  eps in absolute value, the matrix is considered
  singular }
label 1;
const eps=1.0e-07;
var MX:array[1..20,1..21] of real;
    i,j,row,col,pivot:integer;
    pivotsize,temp,coeff:real;
begin
  { copy A and B into augmented matrix }
  for i:=1 to m do
    begin
      for j:=1 to m do
        MX[i,j]:=A[i,j];
      MX[i,m+1]:=B[i]
    end;

  for row:=1 to m do
    begin
      { find pivot }
      pivot:=row;
      pivotsize:=abs(MX[row,row]);
      for i:=row+1 to m do
        if abs(MX[i,row])>pivotsize
          then begin
                 pivot:=i;
                 pivotsize:=abs(MX[pivot,row])
               end;

      { check that pivot is sufficiently large }
      if pivotsize<eps
        then begin
               writeln(' error in SOLVE');
               writeln(' matrix illconditioned');
               goto 1
             end;

      { interchange rows to pivot }
      for col:=row to m+1 do
        begin
          temp:=MX[row,col];
          MX[row,col]:=MX[pivot,col];
          MX[pivot,col]:=temp
        end;

      { reduce to triangular form }
      for i:=row+1 to m do
        begin
          coeff:=MX[i,row]/MX[row,row];
          for j:=row to m+1 do
            MX[i,j]:=MX[i,j]-coeff*MX[row,j]
        end
    end;

  { find solution by backsubstitution }
  for i:=m downto 1 do
    begin
      temp:=MX[i,m+1];
      for j:=m downto i+1 do
        temp:=temp-MX[i,j]*X[j];
      X[i]:=temp/MX[i,i]
    end;
1:end;

procedure VOLT2(h:real;N:integer);
{ procedure for solving a system of m nonlinear
  Volterra equations of the second kind by the
  trapezoidal method. Details are given in Section 12.1 }
label 1;
const tolerance=1.0e-8;
      maxit=5;
var JACOBIAN:MATRIX;
    chi,resid,correction:VECTOR;
    i,j,n,iteration:integer;
    corrnorm:real;
begin
  t[0]:=0;
  for i:=1 to m do
    F[0][i]:=g(i,t[0]);
  for n:=1 to N do
    begin
      t[n]:=t[n-1]+h;
      for i:=1 to m do
        begin
          { compute chi in equation (12.5) }
          chi[i]:=g(i,t[n])+0.5*h*K(i,t[n],t[0],F[0]);
          for j:=1 to n-1 do
            chi[i]:=chi[i]+h*K(i,t[n],t[j],F[j]);
        end;
      iteration:=0;
      for i:=1 to m do
        F[n][i]:=F[n-1][i];
      { Newton iteration to solve for F[n] }
      repeat
        iteration:=iteration+1;
        for i:=1 to m do
          for j:=1 to m do
            if i=j
              then JACOBIAN[i,j]:=1-0.5*h*KDIFF(i,j,t[n],t[n],F[n])
              else JACOBIAN[i,j]:=-0.5*h*KDIFF(i,j,t[n],t[n],F[n]);
        for i:=1 to m do
          resid[i]:=F[n][i]-0.5*h*K(i,t[n],t[n],F[n])-chi[i];
        SOLVE(JACOBIAN,resid,correction);
        { compute norm of correction }
        corrnorm:=0;
        for i:=1 to m do
          if abs(correction[i])>corrnorm
            then corrnorm:=abs(correction[i]);
        for i:=1 to m do
          F[n][i]:=F[n][i]-correction[i]
      until (corrnorm<=tolerance) or (iteration=maxit);
      { test for convergence- nonconvergence is
        recognized if after maxit iteration the
        norm of the correction is still larger
        than tolerance }
      if corrnorm>tolerance
        then begin
               writeln(' error in VOLT2');
               writeln(' Newton method does not converge');
               goto 1
             end
    end;
1:end;
```

FIG. 12.2. *Pascal procedure* VOLT2.

12.2. The product trapezoidal method for a system of the second kind.

We next consider the system

$$\mathbf{f}(t) = \mathbf{g}(t) + \int_0^t p(t,s)\mathbf{K}(t,s,\mathbf{f}(s))\,ds, \qquad (12.10)$$

using the product trapezoidal method described in § 8.3. Taking exactly the same approach and using the notation in § 12.1, we obtain the scheme

$$\mathbf{F}_n^{(i+1)} = \mathbf{F}_n^{(i)} - J^{-1}(\mathbf{F}_n^{(i)})\{\mathbf{F}_n^{(i)} - \beta_{nn}\mathbf{K}(t_n, t_n, \mathbf{F}_n^{(i)}) - \boldsymbol{\chi}_n\}, \qquad (12.11)$$

where

$$\boldsymbol{\chi}_n = \mathbf{g}(t_n) + \alpha_{n1}\mathbf{K}(t_n, t_0, \mathbf{F}_0) + \sum_{j=1}^{n-1}(\alpha_{n,j+1} + \beta_{nj})K(t_n, t_j, \mathbf{F}_j), \qquad (12.12)$$

and $J(\mathbf{u})$ has components

$$[J(\mathbf{u})]_{ij} = \delta_{ij} - \beta_{nn}\frac{\partial K_i(t_n, t_n, \mathbf{u})}{\partial u_j}. \qquad (12.13)$$

This method is implemented in the procedure VOLT2PROD. Its usage is identical to that of VOLT2, except that two additional real functions ALPHA and BETA have to be supplied. With i, j integer and h real, these are to be defined so that

$$\text{ALPHA}(i, j, h) = \frac{1}{h}\int_{t_{j-1}}^{t_j}(t_j - s)p(t_i, s)\,ds, \qquad (12.14)$$

$$\text{BETA}(i, j, h) = \frac{1}{h}\int_{t_{j-1}}^{t_j}(s - t_{j-1})p(t_i, s)\,ds. \qquad (12.15)$$

A listing of VOLT2PROD is given in Fig. 12.3. The code utilizes the procedure SOLVE given in the listing of VOLT2, Fig. 12.2.

```
procedure VOLT2PROD(h:real;N:integer);
{ procedure for solving a system of m nonlinear
  Volterra equations by the product trapezoidal
  method. For details, see Section 12.2 }
label 1;
const tolerance=1.0e-8;
      maxit=8;
var JACOBIAN:MATRIX;
    chi,resid,correction:VECTOR;
    i,j,n,iteration:integer;
    corrnorm:real;
begin
  t[0]:=0;
  for i:=1 to m do
    F[0][i]:=g(i,t[0]);
  for n:=1 to N do
  begin
    t[n]:=t[n-1]+h;
    for i:=1 to m do
    begin
      { compute chi in equation (12.12) }
      chi[i]:=g(i,t[n])+ALPHA(n,1,h)
              *K(i,t[n],t[0],F[0]);
      for j:=1 to n-1 do
        chi[i]:=chi[i]+(ALPHA(n,j+1,h)+BETA(n,j,h))
                *K(i,t[n],t[j],F[j]);
    end;
    iteration:=0;
    for i:=1 to m do
      F[n][i]:=F[n-1][i];
    { Newton iteration for solving for F[n] }
    repeat
      iteration:=iteration+1;
      for i:=1 to m do
        for j:=1 to m do
          if i=j
            then JACOBIAN[i,j]:=1-BETA(n,n,h)
                                *KDIFF(i,j,t[n],t[n],F[n])
            else JACOBIAN[i,j]:=-BETA(n,n,h)
                                *KDIFF(i,j,t[n],t[n],F[n]);
      for i:=1 to m do
        resid[i]:=F[n][i]-BETA(n,n,h)*K(i,t[n],t[n],F[n])-chi[i];
      SOLVE(JACOBIAN,resid,correction);
      corrnorm:=0;
      { compute norm of correction }
      for i:=1 to m do
        if abs(correction[i])>corrnorm
          then corrnorm:=abs(correction[i]);
      for i:=1 to m do
        F[n][i]:=F[n][i]-correction[i]
    until (corrnorm<=tolerance) or (iteration=maxit);
    { test for convergence- nonconvergence is
      recognized if after maxit iterations the
      norm of the correction is still larger than
      tolerance }
    if corrnorm>tolerance
      then begin
        writeln(' error in VOLT2PROD');
        writeln(' Newton method does not converge');
        goto 1
      end
  end;
1:end;
```

FIG. 12.3. *Pascal procedure* VOLT2PROD.

12.3. The midpoint method for systems of the first kind.

For equations of the first kind we take the system

$$\int_0^t \{k_{11}(t,s)f_1(s) + k_{12}(t,s)f_2(s) + \cdots + k_{1m}(t,s)f_m(s)\}\, ds = g_1(t),$$
$$\vdots \qquad\qquad\qquad\qquad\qquad\qquad\qquad (12.16)$$
$$\int_0^t \{k_{m1}(t,s)f_1(s) + k_{m2}(t,s)f_2(s) + \cdots + k_{mm}(t,s)f_m(s)\}\, ds = g_m(t).$$

If we adapt the midpoint method described in Chapter 9 to systems, we are led to the approximating equations

$$K_{nn} \mathbf{F}_{n-1/2} = \frac{\mathbf{g}(t_n)}{h} - \sum_{j=1}^{n-1} K_{nj} \mathbf{F}_{j-1/2} \qquad (12.17)$$

where K_{nj} is an $m \times m$ matrix

$$K_{nj} = \begin{pmatrix} k_{11}(t_n, t_{j-1/2}) & \cdots & k_{1m}(t_n, t_{j-1/2}) \\ \vdots & & \vdots \\ k_{m1}(t_n, t_{j-1/2}) & \cdots & k_{mm}(t_n, t_{j-1/2}) \end{pmatrix}. \qquad (12.18)$$

The procedure VOLT1 computes the solution of (12.17) for $n = 1, 2, \ldots, N$. The usage is essentially identical with that of VOLT2. However, in place of K and KDIFF, a real function k must be provided, such that, with i, j integer, and t, s, real, $k(i, j, t, s)$ has the value of $k_{ij}(t,s)$ in (12.16).

The results produced by VOLT1 are stored in t and F. In the array t, t[0] contains t_0, t[1] contains t_1, and so on. However, the results are stored in F such that F[0] contains the vector $\mathbf{F}_{1/2}$, F[1] the vector $\mathbf{F}_{3/2}$, and so on.

```
procedure VOLT1(h:real;N:integer);
{ procedure for solving a system of m linear
  Volterra equation of the first kind by the
  midpoint method. For details, see Section 12.3 }
var KMX:MATRIX;
    rightside:VECTOR;
    i,j,n,r:integer;
begin
  t[0]:=0;
  for n:=1 to N do
    begin
    t[n]:=t[n-1]+h;
    for i:=1 to m do
      begin
      { compute right side of equation (12.17) }
      rightside[i]:=g(i,t[n]);
      for j:=1 to n-1 do
        for r:= 1 to m do
          rightside[i]:=rightside[i]-
                    h*k(i,r,t[n],t[j]-h/2)*F[j-1][r];
      end;
    for i:=1 to m do
      for j:=1 to m do
        KMX[i,j]:=h*k(i,j,t[n],t[n]-h/2);
    SOLVE(KMX,rightside,F[n-1])
    end
end;
```

FIG. 12.4. *Pascal procedure* VOLT1.

A listing of VOLT1 is given in Fig. 12.4. The code utilizes the procedure SOLVE given in the listing of VOLT2, Fig. 12.2.

12.4. The product midpoint method for a system of the first kind. As the final case we take

$$\int_0^t p(t,s)\{k_{11}(t,s)f_1(s)+\cdots+k_{1m}(t,s)f_m(s)\}\,ds = g_1(t),$$
$$\vdots \qquad (12.19)$$
$$\int_0^t p(t,s)\{k_{m1}(t,s)f_1(s)+\cdots+k_{mm}(t,s)f_m(s)\}\,ds = g_m(t).$$

If the product midpoint method (10.8) is adapted to a system of equations, we get the scheme

$$w_{n,n-1}\mathbf{K}_{nn}\mathbf{F}_{n-1/2} = \mathbf{g}(t_n) - \sum_{j=1}^{n-1} w_{n,j-1}\mathbf{K}_{nj}\mathbf{F}_{j-1/2}, \qquad (12.20)$$

where

$$w_{nj} = \int_{t_j}^{t_{j+1}} p(t_n, s)\,ds. \qquad (12.21)$$

The rest of the notation is the same as in § 12.3. Therefore, only minor changes are required to convert VOLT1 into the product midpoint procedure VOLT1PROD. The usage of VOLT1PROD is identical to VOLT1. An additional function w, with integer parameters n, j and real parameter h, has to be provided so that

$$w(n,j,h) = \int_{t_j}^{t_{j+1}} p(t_n, s)\,ds. \qquad (12.22)$$

```
procedure VOLT1PROD(h:real;N:integer);
{ procedure for solving a system of m linear
Volterra equations of the first kind by the
product midpoint method. For details, see
Section 12.4 }
var KMX:MATRIX;
    rightside:VECTOR;
    i,j,n,r:integer;
begin
  t[0]:=0;
  for n:=1 to N do
    begin
      t[n]:=t[n-1]+h;
      for i:=1 to m do
        begin
          { compute right side of equation (12.20) }
          rightside[i]:=g(i,t[n]);
          for j:=1 to n-1 do
            for r:= 1 to m do
              rightside[i]:=rightside[i]-
                w(n,j-1,h)*k(i,r,t[n],t[j]-h/2)*F[j-1][r];
        end;
      for i:=1 to m do
        for j:=1 to m do
          KMX[i,j]:=w(n,n-1,h)*k(i,j,t[n],t[n]-h/2);
      SOLVE(KMX,rightside,F[n-1])
    end
end;
```

FIG. 12.5. *Pascal procedure* VOLT1PROD.

A listing of VOLT1PROD is given in Fig. 12.5. The code utilizes the procedure SOLVE given in the listing of VOLT2, Fig. 12.2.

Notes on Chapter 12. The article by Miller [188] mentions several programs, including Algol procedures by Pouzet [212] and Rumyantsev [217]. A quite sophisticated Fortran program is given in Logan's thesis [172], but unfortunately, it has not been formally published and consequently is not readily available. A more accessible reference is the Fortran program for a linear equation of the second kind given in Churchhouse [67]. A recently published paper by Delves et al. [85] contains some ALGOL 68 modules, essentially constituting a higher level language for the solution of integral equations. These modules are aimed primarily at the solution of nonstandard equations where the standard methods fail.

Chapter 13

Case Studies

The study of numerical methods provides a general framework for understanding the behavior of approximating algorithms. It tells us under what conditions the methods will work and gives some insight into their relative merits. However, in practice, various questions arise which are not immediately answered by the theory. The problem at hand may be somewhat different from the ones used in the analysis, the assumptions of certain theorems may be violated, or the conclusions inadequate. When this happens, the general ideas incorporated into the theory will have to be modified to deal with specific questions. Often these modifications cannot be rigorously defended. Nevertheless, the insight gained from the theory allows us to obtain approximate results with a high degree of confidence.

To elaborate on this, we study three problems with a view towards dealing with some of these practical questions. These problems are still somewhat simpler than what one can expect to get in practice, but they do have some realistic features and raise questions which are common.

13.1. Estimating errors in the approximation. For our first case study we take a more or less arbitrary example with no relation to any particular practical application. We use it only to study the question of the accuracy of the solution.

Consider the system

$$f_1(t) = t + \int_0^t \left\{ \frac{1}{1+f_1^2(s)} + \cos{(t-s)} f_2(s) \right\} ds, \qquad (13.1)$$

$$f_2(t) = 1 + \int_0^t \left\{ \sin{(t-s)} f_1(s) - e^{s-t} f_2(s) \right\} ds. \qquad (13.2)$$

This set of equations differs in one respect from the various examples throughout the book—we do not know its exact solution. When studying algorithms we normally use simple examples with known results. This is done to demonstrate that the expected behavior of the method is indeed realized (or to discover features not brought to light by the theory). In actual practice the answers are unknown and the problem is to compute an approximation, with some assurance of the accuracy of the results.

The discussion in Chapter 7, in particular (7.19), raises the possibility of computing explicit and rigorous error bounds. If such bounds were usable we could obtain a guarantee that the error would be within a certain tolerance. Unfortunately, in most actual situations, this turns out to be impractical. Rigorous error bounds are not useful even for such simple cases as the system (13.1) and (13.2).

To use (7.19), it is necessary to bound the consistency error $\delta(h, t)$. Since this involves the unknown solution, it is necessary to determine some of its properties, such as bounds on its various derivatives. This may prove impossible or at least impractical. (The reader might try to compute bounds for **f** and its first two derivatives in the system (13.1), (13.2).) At best, we must be satisfied with rather crude answers. A second problem with (7.19) is that it tends to give unrealistically large bounds for the error. The situation is quite common in numerical analysis; rigorous error bounds are seldom useful as a practical means for determining the accuracy of the results.

Anyone experienced in numerical computation will have a quite pragmatical way of dealing with the difficulty. One simply computes several solutions, using increasingly smaller stepsizes, until the answers have "settled down" to the desired accuracy. In Table 13.1 we give the computed results for $f_1(t)$ using the procedure VOLT2 with stepsizes $h = 0.2$ and $h = 0.1$. An examination of the entries will lead us to conclude that the answers for $h = 0.1$ are accurate to about 0.005.

This approach to estimating accuracy appears rather naive, but one rarely can do much better. But we can use our knowledge of what to expect from the method to increase our confidence in this claim and perhaps improve the results slightly.

The method used has order of convergence two for sufficiently smooth kernels and solutions. The system (13.1), (13.2) is certainly well behaved

TABLE 13.1
Approximation to $f_1(t)$ in (13.1), (13.2), using procedure VOLT2.

t	$h = 0.2$	$h = 0.1$
0.2	0.55775	0.56160
0.4	1.03113	1.03412
0.6	1.43660	1.43880
0.8	1.80150	1.80560
1.0	2.14443	2.14889

and it is not hard to show that f_1 and f_2 are infinitely differentiable. Therefore, we can expect second order convergence. The trapezoidal method has a repetition factor one and consequently there is an asymptotic expansion for the discretization error. If we write $Y(t, h)$ for the approximation to $f(t)$ with stepsize h, then we expect that

$$Y(t, h) \simeq f(t) + h^2 e(t) \tag{13.3}$$

and

$$Y\left(t, \frac{h}{2}\right) \simeq f(t) + \frac{h^2}{4} e(t). \tag{13.4}$$

Therefore, the error in the results computed with $h/2$ is approximately

$$\frac{h^2}{4} e(t) \simeq \frac{Y(t, h) - Y(t, h/2)}{3}. \tag{13.5}$$

In other words, the error of the best results is about one-third of the tabular difference. For Table 3.1, this implies that the numbers in the column $h = 0.1$ have an apparent error slightly larger than 0.001.

The arguments leading to (13.5) assume that h is so small that higher order terms in the asymptotic expansion of the error can be neglected. What "small" means in this context is very hard to say. If the stage has not been reached where higher orders of h are negligible, our conclusions may be incorrect.

To reduce the chance for a mistaken conclusion, we can compute the results for yet another stepsize. With three answers available at each point, we can use Aitken's method to estimate the order of convergence by

$$p \simeq \log_2 \frac{Y(t, h/4) - Y(t, h/2)}{Y(t, h/2) - Y(t, h)}. \tag{13.6}$$

If (13.6) yields a number close to two, we can be confident that (13.3) holds reasonably well. In that case, the results can be further improved by the extrapolation

$$Y^E = \frac{2^p Y(t, h/4) - Y(t, h/2)}{2^p - 1}. \tag{13.7}$$

A complete set of results for these computations is given in Table 13.2.

The estimated order of convergence in Table 13.2 is very close to the expected value of two. The maximum difference in results of columns $h = 0.1$ and $h = 0.05$ is about 0.00120 so that the error in column $h = 0.05$ can be expected to be about

$$\frac{0.00120}{3} = 0.0004.$$

TABLE 13.2
Results for $f_1(t)$ in (13.1), (13.2) using VOLT2, with estimated order of convergence and extrapolated values.

t	$h=0.2$	$h=0.1$	$h=0.05$	p	Y^E
0.2	0.55775	0.56160	0.56252	2.06	0.56281
0.4	1.03113	1.03350	1.03412	1.93	1.03434
0.6	1.43660	1.43880	1.43937	1.95	1.43957
0.8	1.80150	1.80460	1.80535	2.05	1.80559
1.0	2.14443	2.14889	2.15004	1.96	2.15044

Actually, since the stepsize is small enough to permit extrapolation, Y^E should be even more accurate. We can therefore with reasonable confidence claim that the column Y^E is an approximation to $f_1(t)$ with an error not much larger than 1×10^{-4}.

13.2. An example from polymer rheology. The equation

$$\mu f'(t) = f^3(t) g(t) + \int_0^t k(t-s) \left\{ \frac{f^3(t)}{f^2(s)} - f(s) \right\} ds \qquad (13.8)$$

models the elongation of a filament of a certain polyethylene which is stretched on the time interval $-\infty < t \le 0$, then released and allowed to undergo elastic recovery for $t > 0$.

While this equation is of the form (11.1), the methods described in Chapter 11 may have to be modified to take into account the form and behavior of the kernel $k(t)$. In many cases, the kernel can be written as

$$k(t) = \sum_{i=1}^m a_i e^{-t/\tau_i}. \qquad (13.9)$$

This suggests an attack on (13.8) through a conversion to a system of differential equations. Using the ideas described in Chapter 1, with some modifications to take care of the nonlinearity, we introduce

$$v_i(t) = \int_0^t e^{-(t-s)/\tau_i} f(s) \, ds, \qquad (13.10)$$

and

$$w_i(t) = \int_0^t e^{-(t-s)/\tau_i} \frac{1}{f^2(s)} \, ds. \qquad (13.11)$$

Then

$$v_i'(t) = f(t) - \frac{1}{\tau_i} v_i(t) \qquad (13.12)$$

and
$$w'_i(t) = \frac{1}{f^2(t)} - \frac{1}{\tau_i} w_i(t). \tag{13.13}$$

Substituting (13.12) and (13.13) into (13.8) gives

$$\mu f'(t) = f^3(t)g(t) + f^3(t) \sum_{i=1}^{m} a_i w_i(t) - \sum_{i=1}^{m} a_i v_i(t). \tag{13.14}$$

Equations (13.12)–(13.14) constitute a system of $2m+1$ ordinary differential equations with given initial conditions

$$w_i(0) = v_i(0) = 0.$$

Since f is the elongation of the filament, its value at $t=0$ must be given as part of the problem.

The difficulty encountered in using (13.12)–(13.14) comes from the actual values in (13.9). In a typical situation, the values of a_i and τ_i range over many orders of magnitude. For example, we might have $a_1 = 10^{-3}$, $\tau_1 = 10^3$ and $a_m = 10^9$, $\tau_m = 10^{-4}$. This makes the kernel very peaked near $t=0$. The resulting differential equations are said to be stiff. Great care must be taken in solving stiff differential equations, although some quite efficient programs are available.

A second approach, necessary when the kernel does not have a simple form like (13.9), is to convert (13.8) into a system of integral equations. If we use

$$z(t) = \mu f'(t), \tag{13.15}$$

then

$$z(t) = f^3(t)g(t) + \int_0^t k(t-s) \left\{ \frac{f^3(t)}{f^2(s)} - f(s) \right\} ds, \tag{13.16}$$

$$f(t) = f(0) + \frac{1}{\mu} \int_0^t z(s)\, ds. \tag{13.17}$$

The presence of $f(t)$ in the integral of (13.16) makes the equation of nonstandard form and the methods we have described are not immediately applicable. A somewhat different arrangement is more satisfactory. We introduce

$$v(t) = \int_0^t k(t-s) \frac{1}{f^2(s)}\, ds, \tag{13.18}$$

$$w(t) = \int_0^t k(t-s) f(s)\, ds, \tag{13.19}$$

then integrate (13.8) to give

$$\mu f(t) = \mu f(0) + \int_0^t f^3(s) g(s) \, ds + \int_0^t \{f^3(s) v(s) - w(s)\} \, ds. \quad (13.20)$$

The system composed of (13.18), (13.19), and (13.20) is now in standard form.

Unfortunately, none of our programs is quite suitable for this problem. VOLT2 would not work well because of the presence of the rapidly varying kernel $k(t-s)$ in (13.18) and (13.19). VOLT2PROD cannot be used because it was written under the assumption that the same factor $p(t, s)$ occurs in all equations. Here $k(t-s)$ is missing in (13.20). This illustrates the point that, when dealing with integral equations, "canned" programs are rarely useful and may have to be modified to suit the particular circumstances.

It would not be particularly difficult to modify VOLT2PROD to take care of this case. What we would need to do is to index the function $p(s, t)$ in (12.10), as well as the corresponding α, β in (8.29) and (8.30), then make the appropriate changes in VOLT2PROD. However, a further complication may arise. In typical cases, $f(t)$ changes rapidly near $t=0$, so that an algorithm with variable stepsizes is needed.

Actually, for (13.8) an attack from first principles is warranted. We introduce the meshpoints $0 = t_0 < t_1 < t_2 < \ldots$, but now allow for unequal stepsizes by dropping the requirement that $t_i - t_{i-1}$ be constant. Integrating (13.8) from t_{n-1} to t_n using the trapezoidal approximation

$$\mu f(t_n) - \mu f(t_{n-1}) \simeq \frac{t_n - t_{n-1}}{2} \left\{ f^3(t_n) g(t_n) + f^3(t_{n-1}) g(t_{n-1}) + H(t_n) + H(t_{n-1}) \right\}$$

$$(13.21)$$

where

$$H(t) = \int_0^t k(t-s) \left\{ \frac{f^3(t)}{f^2(s)} - f(s) \right\} ds. \quad (13.22)$$

The integral in (13.22) is now approximated by product integration using a piecewise linear approximation to the function in the curly brackets. This gives

$$H(t_n) \simeq \sum_{j=0}^n w_{nj} \left\{ \frac{f^3(t_n)}{f^2(t_j)} - f(t_j) \right\}, \quad (13.23)$$

where

$$w_{nj} = (1 - \delta_{0j}) \int_{t_{j-1}}^{t_j} k(t_n - s) \frac{s - t_{j-1}}{t_j - t_{j-1}} \, ds + (1 - \delta_{nj}) \int_{t_j}^{t_{j+1}} k(t_n - s) \frac{t_{j+1} - s}{t_{j+1} - t_j} \, ds.$$

$$(13.24)$$

Finally, we replace $f(t_n)$ by its approximation F_n and write (13.21) as an equality. Thus

$$\mu F_n - \mu F_{n-1} = \frac{t_n - t_{n-1}}{2} \left\{ F_n^3 g(t_n) + F_{n-1}^3 g(t_{n-1}) \right.$$
$$\left. + \sum_{j=0}^{n} w_{nj} \frac{F_n^3 - F_j^3}{F_j^2} + \sum_{j=0}^{n-1} w_{n-1,j} \frac{F_{n-1}^3 - F_j^3}{F_j^2} \right\}. \quad (13.25)$$

Equation (13.25) is a cubic equation for F_n and is easily solved in a step by step fashion for $F_1, F_2, \ldots$. Numerical experiments with the system (13.12)–(13.14) and method (13.25) yielded consistent results. Both are viable numerical algorithms, provided some care is taken to take into account the rapid changes near $t = 0$.

13.3. Solving an equation of the first kind in the presence of large data errors. For this study we consider the solution of the equation

$$\int_0^t \frac{e^{-(t-s)}}{1+t-s} f(s) \, ds = g(t), \quad (13.26)$$

where $g(t)$ is not known in functional form, but rather as a table of measured values. The data with which we will be working is given in Table 13.3, where g_i denotes the observed value for $g(t)$ at $t = t_i$.

The data in Table 13.3 were generated to simulate a typical experimental situation. In such cases one can expect that experimental errors and uncertainties will play a role in the solution. When the entries in Table 13.3 are plotted (Fig. 13.1), we see that the values do not vary smoothly, but seem to oscillate about a smooth function with an error of order of magnitude 0.05 or so. This should alert us to potential trouble.

If we solve (13.26) using the midpoint method with $h = 0.05$, we expect an error magnification by a factor $1/h$, so this approach cannot be considered very promising. The results obtained by a direct application of the midpoint method are plotted in Fig. 13.2. They show the typical oscillatory behavior generally encountered in the numerical solution. Little useful information can be extracted from these results.

To obtain more acceptable results several methods of attack can be chosen. One of the simplest is to smooth the given data, that is, construct a smooth function $\tilde{g}(t)$ such that

$$\tilde{g}(t_i) \simeq g_i, \quad i = 1, 2, \ldots, n,$$

but where $\tilde{g}(t)$ comes from a properly chosen class of functions. One of the most common smoothing techniques is to write $\tilde{g}(t)$ as a linear combination

$$\tilde{g}(t) = c_1 \psi_1(t) + c_2 \psi_2(t) + \cdots + c_m \psi_m(t), \quad (13.27)$$

TABLE 13.3
Original data for (13.26) and smoothed results.

t_i	g_i	Smoothed g_i with $M=4$	Smoothed g_i with $M=5$
0.05	0.059	0.050	0.045
0.10	0.105	0.098	0.097
0.15	0.119	0.145	0.150
0.20	0.211	0.191	0.197
0.25	0.211	0.233	0.236
0.30	0.299	0.271	0.269
0.35	0.315	0.305	0.299
0.40	0.320	0.337	0.330
0.45	0.342	0.367	0.365
0.50	0.413	0.400	0.401
0.55	0.420	0.430	0.437
0.60	0.481	0.462	0.469
0.65	0.518	0.492	0.495
0.70	0.492	0.519	0.516
0.75	0.538	0.541	0.535
0.80	0.551	0.559	0.555
0.85	0.593	0.576	0.576
0.90	0.583	0.598	0.602
0.95	0.643	0.630	0.634
1.00	0.672	0.676	0.674

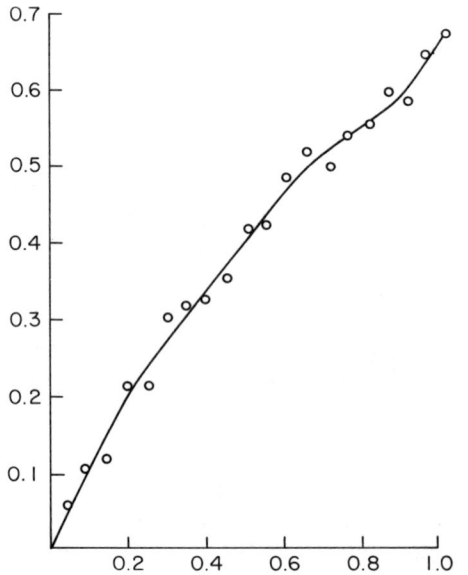

FIG. 13.1. *Original and smoothed data for* (13.26).

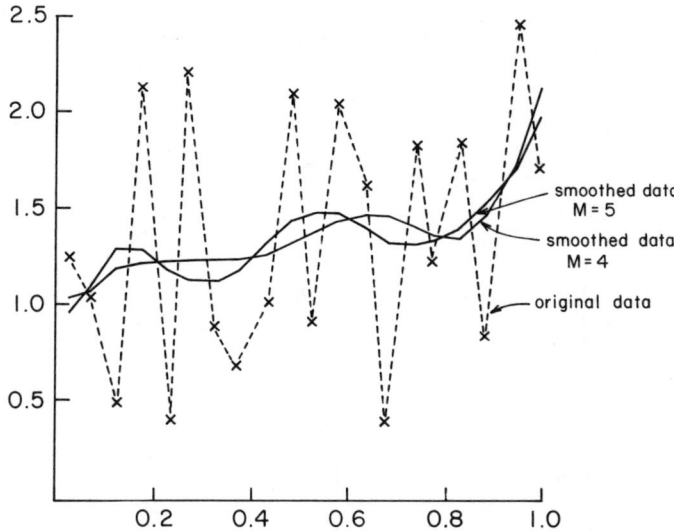

FIG. 13.2. *Solution of* (13.26) *by the midpoint method with original and smoothed data.*

where the $\psi_i(t)$ are a set of expansion functions. The coefficients c_i can be determined in various ways, for example, by the method of least squares. In this method, the coefficients are chosen so that

$$\sum_{i=1}^{n} (\tilde{g}(t_i) - g_i)^2$$

is minimized. The minimization leads to the well-known normal equations. The most frequently used expansion functions are polynomials, trigonometric functions, and splines.

In our study we computed the least squares cubic spline approximation by choosing for ψ_i the so-called B-splines, with some modifications to constrain the solution so that $\tilde{g}(0) = 0$. For this case we used splines with M equidistant knots on the interval $[0, 1]$. Table 13.4 and Fig. 13.1 show the smoothed data values using $M = 4$ and $M = 5$. The two sets of smoothed values are so close together that they are virtually indistinguishable on the graph.

When we use the smoothed values for the midpoint method with $h = 0.05$, we obtain results which are better than before, but still show some oscillations. While the results for $M = 4$ and $M = 5$ agree on the general shape of the solution, they differ by about 10% (Fig. 13.2). Perhaps, this should not be too surprising. The uncertainties in the original data were of that order of magnitude, so that a similar error in the result is not unreasonable. Still, the

TABLE 13.4
Results for (13.26) by the midpoint method, using original and smoothed data.

t_i	Original data	Smoothed data with $M=4$	Smoothed data with $M=5$
0.025	1.24	1.03	0.95
0.075	1.08	1.13	1.17
0.125	0.49	1.19	1.28
0.175	2.16	1.23	1.29
0.225	0.40	1.24	1.20
0.275	2.25	1.23	1.14
0.325	0.90	1.23	1.14
0.375	0.69	1.24	1.21
0.425	1.05	1.27	1.34
0.475	2.12	1.33	1.44
0.525	0.90	1.40	1.48
0.575	2.04	1.45	1.47
0.625	1.65	1.47	1.40
0.675	0.39	1.46	1.35
0.725	1.85	1.41	1.34
0.775	1.24	1.35	1.36
0.825	1.87	1.36	1.44
0.875	0.85	1.48	1.56
0.925	2.29	1.73	1.74
0.975	1.75	2.10	1.97

results illustrate the error magnification inherent in the solution of Volterra equations of the first kind.

Of course, all three functions shown in Fig. 13.2 are "solutions" to (13.26), because for each, the residuals

$$\left| \int_0^t \frac{e^{-(t_i-s)}}{1+t_i-s} \hat{f}(s)\, ds - g_i \right|$$

are small enough to be in the range of the expected fluctuations in g_i. But no doubt, we would reject the result for the unsmoothed data as too erratic and giving no information. To choose between the results for $M=4$ and $M=5$ is a little harder; they can both be considered plausible answers. Usually, to define "plausible" means bringing in some additional information, such as the statistical properties of the fluctuations in g_i, the expected shape of $\hat{f}(t)$, and so on. Without such additional information we cannot make a meaningful choice between the cases $M=4$ and $M=5$, or the results produced by some other smoothing method. If we had to make the choice without any further information, we would probably choose $M=4$ on the grounds that it gives a somewhat less oscillatory answer than $M=5$. In fact, many people would consider both cases $M = 4$ and $M = 5$ as somewhat unsatisfactory and may look for smoother functions which also approximately account for the observed data. This can be done by using still fewer B-splines or by constraining the approximating function further. In this way, a number of different, but plausible answers can be obtained.

Thus, while it is not difficult to produce a plausible solution to (13.26), it is very difficult to assign much meaning to it, in particular, how it is related to other plausible solutions. This is one of the basic, unsolved problems in the study of ill-posed equations.

Notes on Chapter 13. The example from polymer rheology used in § 13.2 is described in Lodge et al. [171].

References

[1] L. F. ABD-ELAL, *An asymptotic expansion for a regularization technique for numerical singular integrals and its application to Volterra integral equations*, J. Inst. Math. Appl., 23 (1979), pp. 291–309.
[2] J. ALBRECHT AND L. COLLATZ, *Numerical Treatment of Integral Equations*, Int. Ser. Num. Math., Vol. 53, Birkhäuser Verlag, Basel, 1980.
[3] A. S. ANDERSSEN AND E. T. WHITE, *Improved numerical methods for Volterra integral equations of the first kind*, Comput. J., 14 (1971), pp. 442–443.
[4] R. S. ANDERSSEN, F. R. DE HOOG, AND R. WEISS, *On the numerical solution of Brownian motion processes*, J. Appl. Prob., 10 (1973), pp. 409–418.
[5] R. S. ANDERSSEN AND P. BLOOMFIELD, *A time series approach to numerical differentiation*, Technometrics, 16 (1974), pp. 69–75.
[6] ———, *Numerical differentiation procedures for nonexact data*, Numer. Math., 22 (1974), pp. 157–182.
[7] R. S. ANDERSSEN and A. J. JAKEMAN, *Product integration for functionals of particle size distribution*, Utilitas Mathematica, 8 (1975), pp. 111–126.
[8] ———, *Abel type integral equations in stereology. II. Computational methods of solution and random spheres approximation*, J. Microscopy, 105 (1975), pp. 135–153.
[9] R. S. ANDERSSEN, *Application and numerical solution of Abel-type integral equations*, Tech. Summary Rept. 1787, Math. Res. Center., Univ. of Wisconsin, Madison, 1977.
[10] ———, *Stable procedures for the inversion of Abel's equation*, J. Inst. Math. Appl., 17 (1976), pp. 329–342.
[11] R. S. ANDERSSEN, F. DE HOOG, AND M. A. LUKAS, eds., *The Application and Numerical Solution of Intergral Equations*, Sijthoff and Noordhoff, Alphen aan den Rijn, 1980.
[12] C. ANDRADE and S. MCKEE, *On optimal high accuracy linear multistep methods for first kind Volterra integral equations*, BIT, 19 (1979), pp. 1–11.
[13] K. E. ATKINSON, *The numerical solution of an Abel integral equation by a product trapezoidal method*, SIAM J. Numer. Anal., 11 (1974), pp. 97–101.
[14] ———, *Existence theorem for Abel integral equations*, SIAM J. Math. Anal., 5 (1974), pp. 729–736.
[15] A. V. BAEV AND V. B. GLASKO, *On solution of the converse kinematic problem of seismics by means of a regularizing algorithm*, USSR Comp. Math. and Math. Phys., 16, 4 (1976), pp. 96–106.
[16] C. T. H. BAKER, *Runge–Kutta Methods for Volterra Integral Equations of the Second Kind*, Lecture Notes in Mathematics 630, Springer–Verlag, Berlin, 1978.
[17] ———, *The Numerical Solution of Integral Equations*, Clarendon Press, Oxford, 1977.

[18] C. T. H. BAKER AND M. S. KEECH, *Stability regions in the numerical treatment of Volterra integral equations*, SIAM J. Numer. Anal., 15 (1978), pp. 394–417.

[19] ———, *Stability analysis of certain Runge–Kutta procedures for Volterra integral equations*, ACM Trans. Math. Software, 4 (1978), pp. 305–315.

[20] C. T. H. BAKER, *Structure of recurrence relations in the study of stability in the numerical treatment of Volterra integral and integro-differential equations*, J. Integral Equations, 2 (1980), pp. 11–29.

[21] C. T. H. BAKER AND J. C. WILKINSON, *Stability analysis of Runge–Kutta methods applied to a basic Volterra integral equation*, J. Austral. Math. Soc. (Ser. B), 22 (1981), pp. 518–540.

[22] C. T. H. BAKER AND G. F. MILLER, eds., *Treatment of Integral Equations by Numerical Methods*, Academic Press, London, 1982.

[23] R. BALASUBRAMANIAN, D. H. NORRIE, AND G. DE VRIES, *The application of the least squares finite element method to Abel's integral equation*. Int. J. Numer. Meth. Eng., 14 (1979), pp. 201–209.

[24] P. R. BEESACK, *Comparison theorems and integral inequalities for Volterra integral equations*, Proc. AMS, 20 (1984), pp. 61–66.

[25] R. BELLMAN, *A Survey of the Mathematical Theory of Time-Lag, Retarded Control, and Hereditary Processes*, Rand Corp., 1954.

[26] R. E. BELLMAN AND K. L. COOKE, *Differential-Difference Equations*, Academic Press, New York, 1963.

[27] R. E. BELLMAN, R. KALABA, AND B. KOTKIN, *Differential approximation applied to the solution of convolution equations*, Math. Comp., 18 (1964), pp. 487–491.

[28] B. A. BELTYUKOV, *An analog of the Runge–Kutta method for solutions of nonlinear Volterra integral equations*, Differential Eq., 1 (1965) pp. 417–426.

[29] M. P. BENSON, *Errors in the numerical quadrature for certain singular integrals and the numerical solution of Abel integral equations*, Ph.D thesis, Univ. Wisconsin, Madison, 1973.

[30] J. M. BOWNDS AND J. CUSHING, *A representation formula for linear Volterra integral equations*, Bull. AMS, 79 (1973), pp. 532–536.

[31] J. M. BOWNDS AND B. WOOD, *On numerically solving nonlinear Volterra integral equations with fewer computations*, SIAM J. Numer. Anal., 13 (1976), pp. 705–719.

[32] J. M. BOWNDS, *On solving weakly singular Volterra equations of the first kind with Galerkin approximation*, Math. Comp., 30 (1976), pp. 747–757.

[33] J. M. BOWNDS AND B. WOOD, *A note on solving Volterra integral equations with convolution kernels*, Appl. Math. Comp., 3 (1977) pp. 305–315.

[34] J. M. BOWNDS, *On an initial-value method for quickly solving Volterra equations: a review*, J. Optim. Theory Appl., 24 (1978), pp. 133–151.

[35] ———, *A modified Galerkin approximation method for Volterra equations with smooth kernels*, Appl. Math. Comp., 4 (1978), pp. 67–79.

[36] ———, *On an initial value method for quickly solving Volterra integral equations*, in Golberg [116], pp. 225–243.

[37] J. M. BOWNDS AND B. WOOD, *A smoothed projection method for singular, nonlinear Volterra equations*, J. Approx. Theory, 25 (1979), pp. 120–141.

[38] J. M. BOWNDS, *A combined recursive collocation and kernel approximation technique for certain singular Volterra integral equations*, J. Integral Equations, 1 (1978), pp. 153–164.

[39] H. W. BRANCA, *The nonlinear Volterra equation of Abel's kind and its numerical treatment*, Computing, 20 (1978), pp. 307–324.

[40] F. BRAUER, *On a nonlinear integral equation for population growth problems*, SIAM J. Math. Anal., 6 (1975), pp. 312–317.

[41] ———, *Constant rate harvesting of a population governed by Volterra integral equations*, J. Math. Anal. Appl., 56 (1976), pp. 18–27.

[42] H. BRUNNER, *The solution of nonlinear Volterra integral equations by piecewise polynomials*, Proc. First Manitoba Conference on Numerical Mathematics, 1971, pp. 65–78.
[43] ———, *The solution of Volterra integral equations of the first kind by piecewise polynomials*, J. Inst. Math. Appl., 12 (1973), pp. 295–302.
[44] ———, *The numerical solution of a class of Abel integral equations by piecewise polynomials*, J. Comp. Phys., 12 (1973), pp. 412–416.
[45] ———, *On the numerical solution of nonlinear Volterra integro-differential equations*, BIT, 13 (1973), pp. 381–390.
[46] ———, *On the numerical solution of a class of Abel integral equations*, Proc. Third Manitoba Conference on Numerical Mathematics, 1973, pp. 105–122.
[47] ———, *Global solution of the generalized Abel integral equation by implicit interpolation*, Math. Comp., 28 (1974), pp. 61–67.
[48] H. BRUNNER AND J. D. LAMBERT, *Stability of numerical methods for Volterra integro-differential equations*, Computing, 12 (1974), pp. 75–89.
[49] H. BRUNNER, *On the approximate solution of first-kind integral equations of the Volterra type*, Computing, 13 (1974), pp. 67–79.
[50] ———, *Projection methods for the approximate solution of integral equations of the first kind*, Proc. Fifth Manitoba Conference on Numerical Mathematics, 1975, pp. 3–23.
[51] H. BRUNNER AND M. D. EVANS, *Piecewise polynomial collocation for Volterra integral equations of the second kind*, J. Inst. Math. Appl., 20 (1977), pp. 415–423.
[52] H. BRUNNER, *Discretization of Volterra integral equations of the first kind*. Math. Comp., 31 (1977), pp. 708–716.
[53] ———, *Discretization of Volterra integral equations of the first kind (II)*, Numer. Math., 30 (1978), pp. 117–138.
[54] ———, *On superconvergence in collocation methods for Abel integral equations*, Proc. Eighth Manitoba Conference on Numerical Mathematics, 1978, pp. 117–128.
[55] ———, *Superconvergence of collocation methods for Volterra integral equations of the first kind*, Computing, 21 (1979), pp. 151–157.
[56] ———, *A note on collocation methods for Volterra integral equations of the first kind*, Computing, 23 (1979), pp. 171–187.
[57] ———, *Superconvergence in collocation and implicit Runge–Kutta methods for Volterra-type integral equations of the second kind*, in Albrecht and Collatz [2], pp. 54–72.
[58] ———, *The application of the variation of constants formula in the numerical analysis of integral and integro-differential equations*, Utilitas Mathematica, 19 (1981), pp. 255–290.
[59] ———, *A survey of recent advances in the numerical solution of Volterra integral and integro-differential equations*, J. Comp. Appl. Math., 8 (1982), pp. 213–229.
[60] H. BRUNNER, E. HAIRER, AND S. P. NORSETT, *Runge–Kutta theory for Volterra equations of the second kind*, Math. Comp., 19 (1982), pp. 147–163.
[61] B. CAHLON, *Numerical solution of nonlinear Volterra integral equations*, J. Comp. Appl. Math., 7 (1981), pp. 121–128.
[62] G. M. CAMPBELL AND J. T. DAY, *The numerical solution of nonlinear Volterra integral equations*, BIT, 10 (1970), pp. 10–19.
[63] ———, *A block-by-block method for the numerical solution of Volterra integral equations*, BIT, 11 (1974), pp. 120–124.
[64] S. H. CHANG AND J. T. DAY, *On the numerical solution of a certain nonlinear integro-differential equation*, J. Comp. Phys., 26 (1978), pp. 162–168.
[65] B. A. CHARTERS AND R. S. STEPLEMAN, *Order of convergence of linear multistep methods for functional differential equations*, SIAM J. Numer. Anal., 12 (1975), pp. 876–886.
[66] J. CHOVER AND P. NEY, *The nonlinear renewal equation*, J. Analyse Math., 21 (1968), pp. 381–413.
[67] R. F. CHURCHHOUSE, ed., *Numerical Methods*, Vol. 3, Handbook of Applied Mathematics, John Wiley, New York, 1981.

[68] J. A. COCHRAN, *The Analysis of Linear Integral Equations*, McGraw-Hill, New York, 1972.
[69] L. COLLATZ, *Numerische Behandlung von Differentialgleichungen*, Springer-Verlag, Berlin, 1955.
[70] C. CORDUNEANU, *Principles of Differential and Integral Equations*, Chelsea, New York, 1971.
[71] ———, *Integral Equations and Stability of Feedback Systems*, Academic Press, New York, 1973.
[72] C. W. CRYER, *Numerical methods for functional equations*, in Delay and Functional Differential Equations, K. Schmitt, ed., Academic Press, New York, 1972.
[73] C. M. DAFERMOS, *An abstract Volterra equation with application to linear viscoelasticity*, J. Differential Equations, 7 (1970), pp. 554–589.
[74] H. T. DAVIS, *The theory of Volterra integral equations of the second kind*, Indiana Univ. Studies 88–90, Bloomington, IN, 1930.
[75] ———, *Introduction to Nonlinear Differential and Integral Equations*, Dover, New York, 1962.
[76] J. T. DAY, *A starting method for solving nonlinear Volterra integral equations*, Math. Comp., 21 (1967), pp. 178–188.
[77] ———, *On the numerical solution of Volterra integral equations*, BIT, 8 (1968), pp. 134–137.
[78] F. DE HOOG AND R. WEISS, *Asymptotic expansions for product integration*, Math. Comp., 27 (1973), pp. 295–306.
[79] ———, *On the solution of Volterra integral equations of the first kind*, Numer. Math., 21 (1973), pp. 22–32.
[80] ———, *High order methods for Volterra integral equations of the first kind*, SIAM J. Numer. Anal., 10 (1973), pp. 647–664.
[81] ———, *On the solution of a Volterra integral equation with a weakly singular kernel*, SIAM J. Math. Anal., 4 (1973), pp. 561–573.
[82] ———, *High order methods for a class of Volterra integral equations with weakly singular kernels*, SIAM J. Numer. Anal., 11 (1974), pp. 1166–1180.
[83] ———, *Implicit Runge–Kutta methods for second kind Volterra integral equations*, Numer. Math., 23 (1975), pp. 199–213.
[84] L. M. DELVES AND J. WALSH, eds., *Numerical Solution of Integral Equations*, Clarendon Press, Oxford, 1974.
[85] L. M. DELVES, L. F. ABD-ELAL, AND J. A. HENDRY, *A set of modules for the solution of integral equations*, Comput. J., 24 (1981), pp. 184–190.
[86] N. DISTEFANO, *A Volterra integral equation in the stability of linear hereditary phenomena*, J. Math. Anal. Appl., 23 (1968), pp. 365–383.
[87] J. DOUGLAS, JR, *Mathematical programming and integral equations*, in PICC [205], pp. 269–274.
[88] H. EDELS, K. HEARNE, AND A. YOUNG, *Numerical solution of the Abel integral equation*, J. Math. and Phys., 41 (1962), pp. 62–75.
[89] P. B. B. EGGERMONT, *Special discretization for the integral equation of image reconstruction and for Abel-type integral equations*, Tech. Rept. MIPG50, Dept. Computer Science, State Univ. of New York at Buffalo, 1981.
[90] ———, *A new analysis of the trapezoidal method for the numerical solution of Abel-type integral equations*, J. Integral Equations, 3 (1981), pp. 317–332.
[91] M. E. A. EL TOM, *Application of spline functions in Volterra integral equations*, J. Inst. Math. Appl., 8 (1971), pp. 354–357.
[92] ———, *Numerical solution of Volterra integral equations by spline functions*, BIT, 13 (1973), pp. 1–7.

[93] ———, *On the numerical stability of spline function approximations to solutions of Volterra integral equations of the second kind*, BIT, 14 (1974), pp. 136–143.

[94] ———, *On spline function approximation to the solution of Volterra integral equations of the first kind*, BIT, 14 (1974), pp. 288–297.

[95] ———, *Spline function approximation to the solution of singular Volterra integral equations of the second kind*, J. Inst. Math. Appl., 14 (1974), pp. 303–309.

[96] ———, *Efficient algorithms for Volterra integral equations of the second kind*, Computing, 14 (1975), pp. 153–166.

[97] ———, *Application of spline functions to systems of Volterra integral equations of the second kind*, J. Inst. Math. Appl., 17 (1976), pp. 295–310.

[98] R. ESPINOSA-MALDONADO, AND G. D. BYRNE, *Solution of linear integral equations by the Gregory method*, BIT, 10 (1970), pp. 457–464.

[99] R. ESSER, *Numerische Behandlung einer Volterraschen Integralgleichung*, Computing, 19 (1978), pp. 269–284.

[100] A. FELDSTEIN AND J. R. SOPKA, *Numerical methods for nonlinear Volterra integro-differential equations*, SIAM J. Numer. Anal., 11 (1974), pp. 826–846.

[101] W. FELLER, *On the integral equation of renewal theory*, Ann. Math. Stat., 12 (1941), pp. 243–267.

[102] H. E. FETTIS, *On the numerical solution of equations of the Abel type*, Math. Comp., 18 (1964), pp. 491–496.

[103] W. T. FORD, *Mathematical programming and integrodifferential equations*, SIAM J. Numer. Anal., 2 (1965), pp. 171–202.

[104] L. FOX AND E. T. GOODWIN, *The numerical solution of nonsingular linear integral equations*, Phil. Trans. Roy. Soc., London, Ser. A, 245 (1953), pp. 501–534.

[105] R. FRATILA, *An Introduction to the Theory of Linear Systems*, Dept. of the Navy, Electronic System Command, Washington, DC, 1977.

[106] A. FRIEDMAN, *On an integral equation of Volterra type*, J. Analyse Math., 11 (1963), pp. 381–413.

[107] L. GAREY, *Implicit methods for Volterra integral equations of the second kind*, Proc. Second Manitoba Conference on Numerical Mathematics, 1972, pp. 167–177.

[108] ———, *Predictor-corrector methods for nonlinear Volterra integral equations of the second kind*, BIT, 12 (1972), pp. 325–333.

[109] ———, *Numerical methods for second kind Volterra equations with singular kernels*, Proc. Fourth Manitoba Conference on Numerical Mathematics, 1974, pp. 253–263.

[110] ———, *The numerical solution of Volterra integral equations with singular kernels*, BIT, 14 (1974), pp. 33–39.

[111] ———, *Block methods for nonlinear Volterra integral equations*, BIT, 15 (1975), pp. 401–408.

[112] ———, *Solving nonlinear second kind Volterra equations by modified increment methods*, SIAM J. Numer. Anal., 12 (1975), pp. 501–508.

[113] C. J. GLADWIN, *Methods of higher order for the numerical solution of first kind Volterra integral equations*, Proc. Second Manitoba Conference on Numerical Mathematics, 1972, pp. 179–193.

[114] C. J. GLADWIN AND R. JELTSCH, *Stability of quadrature rule methods for first kind Volterra integral equations*, BIT, 14 (1974), pp. 144–151.

[115] C. J. GLADWIN, *Quadrature rule methods for Volterra integral equations of the first kind*, Math. Comp., 33 (1979), pp. 705–716.

[116] M. A. GOLBERG, ed., *Solution Methods for Integral Equations*, Plenum, New York, 1978.

[117] ———, *On a method of Bownds for solving Volterra integral equations*, in Golberg [116], pp. 245–256.

[118] A. GOLDFINE, *Taylor series methods for the solution of Volterra integral and integro-differential equations*, Math. Comp., 31 (1977), pp. 691–707.

[119] A. GOLDMAN AND W. VISSCHER, *Application of integral equations in particle size statistics*, in Golberg [116], pp. 169–182.
[120] P. L. GOLDSMITH, *The calculation of true particle size distributions from the sizes observed in a thin slice*, Brit. J. Appl. Phys., 18 (1967), pp. 813–830.
[121] I. S. GRADSHTEYN AND I. M. RYZHIK, *Tables of Integrals, Series, and Products*, Academic Press, New York, 1965.
[122] J. A. GUZEK AND G. A. KEMPER, *A new error analysis for the cubic spline approximate solution of a class of Volterra integro-differential equations*, Math. Comp., 27 (1973), pp. 563–570.
[123] G. H. HARDY, *Divergent Series*, Clarendon Press, Oxford, 1949.
[124] D. R. HILL, *A new class of one-step methods for the solution of Volterra functional differential equations*, BIT, 14 (1974), pp. 298–305.
[125] J. HILZMAN, *Error bounds for an approximate solution to the Volterra integral equation*, Pacific J. Math., 10 (1960), pp. 203–207.
[126] W. HOCK, *Asymptotic expansion for multistep methods applied to nonlinear Volterra integral equations of the second kind*, Numer. Math., 33 (1979), pp. 77–100.
[127] ———, *An extrapolation method with step size control for nonlinear Volterra integral equations*, Numer. Math., 38 (1981), pp. 155–178.
[128] P. A. HOLYHEAD, S. MCKEE, AND P. J. TAYLOR, *Multistep methods for solving linear Volterra integral equations of the first kind*, SIAM J. Numer. Anal., 12 (1975), pp. 698–711.
[129] P. A. HOLYHEAD AND S. MCKEE, *Stability and convergence of multistep methods for linear Volterra integral equations of the first kind*, SIAM J. Numer. Anal., 13 (1976), pp. 269–292.
[130] I. HOPKINS AND R. HAMMING, *On creep and relaxation*, J. Appl. Phys., 28 (1957), pp. 906–909.
[131] P. J. VAN DER HOUWEN, *Convergence and stability results in Runge–Kutta type methods for Volterra integral equations*, BIT, 20 (1980), pp. 375–377.
[132] P. J. VAN DER HOUWEN AND P. H. M. WOLKENFELT, *On the stability of multistep formulas for Volterra integral equations of the second kind*, Computing, 24 (1980), pp. 341–347.
[133] P. J. VAN DER HOUWEN AND H. J. J. TE RIELE, *Backward differentiation type formulas for Volterra integral equations of the second kind*, Numer. Math., 37 (1981), pp. 205–217.
[134] P. J. VAN DER HOUWEN, P. H. M. WOLKENFELT, AND C. T. H. BAKER, *Convergence and stability analysis for modified Runge–Kutta methods in the numerical treatment of second kind Volterra equations*, IMA J. Numer. Anal., 1 (1981), pp. 303–328.
[135] A. HUBER, *Eine Näherungsmethode zur Auflösung Volterrascher Integralgleichungen*, Monatsh. f. Math. Phys., 47 (1938), pp. 240–246.
[136] H. S. HUNG, *The numerical solution of differential and integral equations by spline functions*, Tech. Summary Rept. 1053, Math. Res. Center, Univ. Wisconsin, Madison, 1970.
[137] ———, *Error analysis of a linear spline method for solving an Abel equation*, Tech. Summary Rept. 1904, Math. Res. Center, Univ. Wisconsin, Madison, 1979.
[138] ———, *A higher order global approximation method for solving an Abel integral equation by quadratic splines*, Tech. Summary Rept. 1921, Math. Res. Center, Univ. Wisconsin, Madison, 1979.
[139] ———, *Spline approximation to the solution of a class of Abel integral equations*, Tech. Summary Rept. 1933, Math. Res. Center, Univ. Wisconsin, Madison, 1979.
[140] K. IGUCHI, *A starting method for solving nonlinear Volterra integral equations of the second kind*, Comm. Assoc. Comp. Mach., 15 (1972), pp. 460–461.
[141] M. K. JAIN AND K. D. SHARMA, *Numerical solution of linear differential equations and Volterra's integral equation using Lobatto quadrature formula*, Comp. J., 10 (1967), pp. 101–107.

[142] A. J. JAKEMAN, *The numerical inversion of Abel type integral equations in stereology*, Ph.D thesis, Australian National University, Canberra, 1975.
[143] A. J. JAKEMAN AND R. S. ANDERSSEN, *Abel type integral equations in stereology I. General discussion*, J. Microscopy, 105 (1975), pp. 121–133.
[144] Z. JACKIEWICZ, *Convergence of multistep methods for Volterra functional differential equations*, Numer. Math., 32 (1979), pp. 307–332.
[145] R. L. JAMES, *The numerical solution of singular Volterra integral equations*, SIAM J. Numer. Anal., 5 (1968), pp. 352–363.
[146] J. G. JONES, *On the numerical solution of convolution equations and systems of such equations*, Math. Comp., 15 (1961), pp. 131–142.
[147] M. S. KEECH, *A third order, semi-explicit method in the numerical solution of first kind Volterra integral equations*, BIT, 17 (1977), pp. 312–320.
[148] G. A. KEMPER, *Linear multistep methods for a class of functional differential equations*, Numer. Math., 19 (1972), pp. 361–372.
[149] ———, *Spline function approximation for solution of functional differential equations*, SIAM J. Numer. Anal., 12 (1975), pp. 73–88.
[150] G. KENT AND J. MAUTZ, *The numerical solution of a Volterra integral equation*, J. Comp. Phys., 3 (1967), pp. 399–415.
[151] D. L. KNIRK, *Accurate and efficient quadrature for Volterra integral equations*, J. Comp. Phys., 21 (1976), pp. 371–399.
[152] M. KOBAYASHI, *On numerical solution of the Volterra integral equation of the second kind by linear multistep methods*, Rep. Stat. Appl. Res., JUSE 13 (1966), pp. 119–139.
[153] ———, *On the numerical solution of the Volterra integral equation of the first kind by the trapezoidal method*, Rep. Stat. Appl. Res., JUSE 14 (1967), pp. 1–14.
[154] E. L. KOSAREV, *The numerical solution of Abel's integral equation*, USSR Comp. Math. and Math. Phys., 13, #6 (1973), pp. 271–277.
[155] G. KOWALEWSKI, *Integralgleichungen*, DeGruyter, Leipzig, 1930.
[156] S. KUMAR, *On a method of Noble for second kind Volterra integral equations*, BIT, 19 (1979), pp. 482–488.
[157] M. LAUDET AND H. OULES, *Sur l'intégration numérique des équations intégrales du type de Volterra*, in PICC [205], pp. 117–121.
[158] V. G. LEVICH, E. M. PODGAETSKI, AND V. YU. FILINOVSKII, *A successive approximation for nonlinear Volterra integral equations of the second kind*, USSR Comp. Math. and Math. Phys., 10, 3 (1978), pp. 138–145.
[159] J. J. LEVIN AND J. A. NOHEL, *On a system of integrodifferential equations occurring in reactor dynamics*, II, Arch. Rat. Mech. Anal., 11 (1962), pp. 210–243.
[160] N. LEVINSON, *A nonlinear Volterra equation arising in the theory of superfluidity*, J. Math. Anal. Appl., 1 (1960), pp. 1–11.
[161] S. P. LIN, *Damped vibration of a string*, J. Fluid Mech., 72 (1975), pp. 787–797.
[162] R. LING, *Integral equations of Volterra type*, J. Math. Anal. Appl., 64 (1973), pp. 381–397.
[163] P. LINZ, *The numerical solution of Volterra integral equations by finite difference methods*, Tech. Summary Rept. 825, Math. Res. Center, Univ. Wisconsin, Madison, 1967.
[164] ———, *A method for solving nonlinear Volterra integral equations of the second kind*, Math. Comp., 23 (1969), pp. 595–600.
[165] ———, *Numerical methods for Volterra integral equations of the first kind*, Comput. J., 12 (1969), pp. 393–397.
[166] ———, *Linear multistep methods for Volterra integrodifferential equations*, J. Assoc. Comput. Mach., 16 (1969), pp. 295–301.
[167] ———, *Numerical methods for Volterra integral equations with singular kernels*, SIAM J. Numer. Anal., 6 (1969), pp. 393–397.

[168] ——, *Product integration methods for Volterra integral equations of the first kind*, BIT, 11 (1971), pp. 413–421.
[169] ——, *A survey of methods for the solution of Volterra integral equations of the first kind*, in Anderssen et al. [11], pp. 183–194.
[170] ——, *The solution of Volterra equations of the first kind in the presence of large uncertainties*, in Baker and Miller [22], pp. 123–130.
[171] A. S. LODGE, J. B. MCLEOD, AND J. A. NOHEL, *A nonlinear perturbed Volterra integrodifferential equation occurring in polymer rheology*, Proc. Roy. Soc. Edinburgh, 80A (1978), pp. 99–137.
[172] J. E. LOGAN, *The approximate solution of Volterra integral equations of the second kind*, Ph.D thesis, Univ. Iowa, Iowa City, 1976.
[173] A. J. LOTKA, *A contribution to the theory of self-moving aggregates with special reference to industrial replacement*, Ann. Math. Stat., 10 (1939), pp. 1–25.
[174] CH. LUBICH, *On the stability of linear multistep methods for Volterra integral equations of the second kind*, in Baker and Miller [22], pp. 233–238.
[175] R. C. MACCAMY AND P. WEISS, *Numerical solution of Volterra integral equations*, Nonlinear Anal., 3 (1979), pp. 677–685.
[176] G. J. MAKINSON AND A. YOUNG, *The stability of solution of differential and integral equations*, in PICC [205], pp. 499–509.
[177] A. MAKROGLOU, *A block-by-block method for Volterra integro-differential equations with weakly singular kernel*, Math. Comp., 37 (1981), pp. 95–99.
[178] L. MALINA, *A-stable methods of higher order for Volterra integral equations*, Aplikace Matematiky, 20 (1975), pp. 336–344.
[179] W. R. MANN AND F. WOLF, *Heat transfer between solids and gases under nonlinear boundary conditions*, Quart. Appl. Math., 9 (1951), pp. 163–184.
[180] J. MATTHYS, *A-stable linear multistep methods for Volterra integro-differential equations*, Numer. Math., 27 (1976), pp. 85–94.
[181] D. F. MAYERS, *Equations of Volterra type*, in Numerical Solution of Ordinary and Partial Differential Equations, L. Fox, ed., Pergamon, Oxford, 1962.
[182] S. MCKEE, *Cyclic multistep methods for solving Volterra integro-differential equations*, SIAM J. Numer. Anal., 16 (1979), pp. 106–114.
[183] ——, *Best convergence rates for linear multistep methods for Volterra first kind equations*, Computing, 21 (1979), pp. 343–358.
[184] ——, *The analysis of a variable step, variable coefficient linear multistep method for solving a singular integrodifferential equation arising in diffusion of discrete particles in a turbulent fluid*, J. Inst. Math. Appl., 23 (1979), pp. 373–388.
[185] S. MCKEE AND H. BRUNNER, *The repetition factor and numerical stability of Volterra integral equations*, J. Comp. Appl. Math., 6 (1980), pp. 329–347.
[186] G. MICULA, *Bermerkungen zur numerischen Behandlung von nichtlinearen Volterraschen Integralgleichungen mit Splines*, Z. Angew. Math. Mech., 56 (1976), pp. T302–304.
[187] S. G. MIKHLIN, *Linear Integral Equations*, Hindustan Publ. Co., Delhi, 1980.
[188] G. F. MILLER, *Provision of library programs for the numerical solution of integral equations*, in Delves and Walsh [84], pp. 247–256.
[189] R. K. MILLER AND A. FELDSTEIN, *Smoothness of solutions of Volterra integral equations with weakly singular kernels*, SIAM J. Math. Anal., 2 (1971), pp. 242–258.
[190] R. K. MILLER, *Nonlinear Volterra Integral Equations*, W. A. Benjamin, Menlo Park, CA, 1971.
[191] G. N. MINERBO AND M. E. LEVY, *Inversion of Abel's integral equation by means of orthogonal polynomials*, SIAM J. Numer. Anal., 6 (1969), pp. 598–616.
[192] W. L. MOCARSKY, *Convergence of step-by-step methods for nonlinear integro-differential equations*, J. Inst. Math. Appl., 8 (1971), pp. 235–239.

[193] O. H. NESTOR AND H. H. OLSEN, *Numerical methods for reducing line and surface probe data*, SIAM Rev., 2 (1960), pp. 200–207.
[194] B. NETA, *Numerical solution of a nonlinear integro-differential equation*, J. Math. Anal. Appl., 89 (1982), pp. 598–611.
[195] A. N. NETRAVALI, *Spline approximation to the solution of the Volterra integral equation of the second kind*, Math. Comp., 27 (1983), pp. 99–106.
[196] B. NOBLE, *The numerical solution of nonlinear integral equations and related topics*, in Nonlinear Integral Equations, P. M. Anselone, ed., Univ. Wisconsin Press, Madison, 1964.
[197] ———, *A bibliography on methods for solving integral equations*, Tech. Summary Rept. 1177, Math. Res. Center, Univ. Wisconsin, Madison, 1971.
[198] ———, *Instability when solving Volterra integral equations of the second kind by multistep methods*, Lecture Notes in Mathematics 109, Springer–Verlag, Berlin, 1969.
[199] J. C. O'NEILL AND G. D. BYRNE, *A starting method for the numerical solution of Volterra's equation of the second kind*, BIT, 8 (1968), pp. 43–47.
[200] H. OULES, *Résolution numérique d'une équation intégrale singulière*, Rev. Francaise Traitment de l'Information, 7 (1964), pp. 117–124.
[201] T. S. PAPATHEODOROU AND M. E. JESANIS, *Collocation methods for Volterra integrodifferential equations with singular kernels*, J. Comp. Appl. Math., 6 (1980), pp. 3–8.
[202] A. G. PETSOULAS, *The approximate solution of Volterra integral equations*, J. Approx. Theory, 14 (1975), pp. 152–159.
[203] G. M. PHILLIPS, *An error estimate for Volterra integral equations*, BIT, 11 (1971), pp. 181–186.
[204] J. R. PHILLIPS, *Some integral equations in geometric probability*, Biometrika, 53 (1966), pp. 365–374.
[205] PICC (Prov. Internat. Comp. Center): *Symposium on the Numerical Treatment of Ordinary Differential Equations, Integral and Integro-differential Equations, Rome 1960*, Birkhauser Verlag, Basel, 1960.
[206] R. PIESSENS AND P. VERBAETEN, *Numerical solution of Abel integral equation*, BIT, 13 (1973), pp. 451–457.
[207] W. POGORZELSKI, *Integral Equations and their Application*, Vol. I, Pergamon, New York, 1966.
[208] G. PORATH, *Störungsrechnung für lineare Volterrasche Integralgleichungen*, Math. Nachr., 37 (1968), pp. 83–98.
[209] V. V. POSPELOV, *The error of Adams type methods for Cauchy–Volterra problem*, USSR Comp. Math. and Math. Phys., 13, 5 (1973), pp. 301–305.
[210] P. POUZET, *Methode d'intégration numérique des équations intégrales et intégrodifferentielles du type de Volterra de seconde espèce. Formules de Runge–Kutta*, in PICC [205], pp. 362–368.
[211] ———, *Etude en vue de leur traitment numérique des équations intégrales de type Volterra*, Rev. Francaise Traitment de l'Information, 6 (1963), pp. 79–112.
[212] ———, *Algorithm de resolution des équations intégrales de type Volterra par des methodes par pas*, Rev. Francaise Traitment de l'Information, 7 (1964), pp. 169–173.
[213] A. PROSPERETTI, *A numerical method for the solution of certain classes of nonlinear Volterra integro-differential equations and integral equations*, Int. J. Numer. Meth. Eng., 11 (1977), pp. 431–438.
[214] J. RADZIUK, *The numerical solution from measurement data of linear integral equations of the first kind*, Int. J. Numer. Meth. Eng., 11 (1977), pp. 729–740.
[215] E. RAKOTCH, *Numerical solution of Volterra integral equations*, Numer. Math., 20 (1973), pp. 271–279.
[216] T. G. ROGERS AND E. H. LEE, *The cylinder problem in viscoelastic stress analysis*, Quart. Appl. Math., 22 (1964), pp. 117–131.

[217] I. A. RUMYANTSEV, *Programme for solving a system of Volterra integral equations (of the second kind)*, USSR Comp. Math. and Math. Phys., 5, 5 (1965), pp. 218–224.
[218] T. SATO, *Sur l'équation intégrale nonlinéaire de Volterra*, Compositio Math., 11 (1953), pp. 271–282.
[219] W. W. SCHMAEDEKE, *Approximate solutions for Volterra integral equations of the first kind*, J. Math. Anal. Appl., 23 (1968), pp. 604–613.
[220] V. O. SERGEEV, *Regularization of Volterra equations of the first kind*, Sov. Math. Dokl., 12 (1971), pp. 501–505.
[221] C. C. SHILEPSKY, *The asymptotic behavior of an integral equation with application to Volterra's population equation*, J. Math. Anal. Appl., 48 (1974), pp. 764–779.
[222] R. SMARZEWSKI AND H. MALINOWSKI, *Numerical solution of a class of Abel integral equations*, J. Inst. Math. Appl., 22 (1978), pp. 159–170.
[223] D. SPOHN, *Sur les formules à pas liés dans l'intégration des équations intégrales du type Volterra*, in Quatrième Congres de Calcul et de Traitment de l'Information, Versailles 1964, Dunod, Paris, 1965.
[224] J. STEINBERG, *Numerical solution of Volterra integral equations*, Numer. Math., 19 (1972), pp. 212–217.
[225] D. R. STOUTEMEYER, *Analytically solving integral equations by using computer algebra*, ACM Trans. Math. Software, 3 (1977), pp. 128–148.
[226] K. E. SWICK, *A nonlinear model for human population dynamics*, SIAM J. Appl. Math., 40 (1981), pp. 266–278.
[227] L. TAVERNINI, *One-step methods for the numerical solution of Volterra functional differential equations*, SIAM J. Numer. Anal., 8 (1971), pp. 786–795.
[228] P. J. TAYLOR, *The solution of Volterra integral equations of the first kind using inverted differentiation formulae*, BIT, 16 (1976), pp. 416–425.
[229] F. G. TRICOMI, *Integral Equations*, Interscience, New York, 1957.
[230] Z. B. TSALYUK, *Volterra integral equations*, J. Soviet Math., 12 (1979), pp. 715–758.
[231] V. VOLTERRA, *Leçons sur la théorie mathématique de la lutte pour la vie*, Gauthier-Villars, Paris, 1931.
[232] ———, *Theory of Functionals and Integral and Integro-differential Equations*, Dover, New York, 1959.
[233] C. WAGNER, *On the numerical solution of Volterra integral equations*, J. Math. and Phys., 32 (1954), pp. 289–303.
[234] F. J. S. WANG, *Asymptotic behavior of some deterministic epidemic models*, SIAM J. Math. Anal., 9 (1978), pp. 529–534.
[235] D. G. WEISS, *Asymptotic behavior of some nonlinear Volterra integral equations*, J. Math. Anal. Appl., 48 (1975), pp. 59–87.
[236] R. WEISS AND R. S. ANDERSSEN, *A product integration method for a class of singular first kind Volterra equations*, Numer. Math., 18 (1972), pp. 442–456.
[237] R. WEISS, *Product integration for the generalized Abel equation*, Math. Comp., 26 (1972), pp. 177–190.
[238] D. WESTREICH AND B. CAHLON, *Numerical solution of Volterra integral equations with continuous and discontinuous terms*, J. Inst. Math. Appl., 26 (1980), pp. 175–186.
[239] D. V. WIDDER, *The Laplace Transform*, Princeton Univ. Press, Princeton, NJ, 1941.
[240] L. F. WIEDERHOLT, *Stability of multistep methods for delay differential equations*, Math. Comp., 30 (1976), pp. 283–290.
[241] K. L. WIGGINS, *Successive approximations to solutions of Volterra integral equations*, J. Approx. Theory, 22 (1978), pp. 340–349.
[242] M. A. WOLFE, *The numerical solution of nonsingular integral and integro-differential equations by iteration with Chebyshev series*, Comput. J., 12 (1969), pp. 193–196.
[243] P. H. M. WOLKENFELT AND P. J. VAN DER HOUWEN, *Analysis of numerical methods for*

second kind Volterra equations by embedding techniques, J. Integral Equations, 3 (1981), pp. 61–82.
[244] P. H. M. WOLKENFELT, *The numerical analysis of reducible quadrature methods for Volterra integral and integro-differential equations*, Akademisch Proefschrift, Mathematisch Centrum, Amsterdam, 1981.
[245] ———, *Linear multistep methods and the construction of quadrature formulae for Volterra integral and integrodifferential equations*, Rept. NW 76/79 (1979), Mathematisch Centrum, Amsterdam.
[246] B. WOOD, *Numerically solving nonlinear Volterra integral equations with fewer computations*, SIAM J. Numer. Anal., 13 (1976), pp. 705–719.
[247] K. YOSIDA, *Lectures on Differential and Integral Equations*, Interscience, New York, 1960.
[248] A. YOUNG, *Approximate product integration*, Proc. Roy. Soc. London, Ser. A, 224 (1954), pp. 552–561.
[249] ———, *The application of approximate product-integration to the numerical solution of integral equations*, Proc. Roy. Soc. London, Ser. A, 224 (1954), pp. 561–573.

Supplementary Bibliography

A list of recent papers on the numerical solution of Volterra equations.

S. AMINI, *On the stability of numerical methods for Volterra integral equations of the second kind*, in Baker and Miller [22], pp. 43–46.
S. AMINI, C. T. H. BAKER, P. J. VAN DER HOUWEN, AND H. J. J. TE RIELE, *Stability analysis of numerical methods for Volterra integral equations with polynomial convolution kernels*, J. Integral Equations, 5 (1983), pp. 73–92.
S. AMINI, *Stability analysis of methods employing reducible rules for Volterra integral equations*, BIT, 23 (1983), pp. 322–328.
R. S. ANDERSSEN AND F. R. DE HOOG, *Application and numerical solution of Abel-type integral equations*, Math. Res. Rept. 7-1982, Australian National University, Canberra, 1982.
C. T. H. BAKER AND J. C. WILKINSON, *On the construction of stability polynomials for modified R-K methods for Volterra integro-differential equations*, in Baker and Miller [22], pp. 33–42.
C. T. H. BAKER, *Stability and structure in numerical methods for Volterra integral equations*, in Baker and Miller [22], pp. 107–122.
J. M. BOWNDS, *Comments on the performance of a FORTRAN subroutine for certain Volterra equations*, in Baker and Miller [22], pp. 163–168.
———, *Theory and performance of a subroutine for solving Volterra integral equations*, Computing, 28 (1982), pp. 317–332.
H. BRUNNER, *On collocation approximations for Volterra equations with weakly singular kernels*, in Baker and Miller [22], pp. 409–420.
———, *Nonpolynomial spline collocation for Volterra equations with weakly singular kernels*, SIAM J. Numer. Anal., 20 (1983), pp. 1106–1119.
———, *Implicit Runge–Kutta methods of optimal order for Volterra integro-differential equations*, Math. Comp., 40 (1984), pp. 95–109.
H. BRUNNER AND H. J. J. TE RIELE, *Volterra type integral equations of the second kind with nonsmooth solutions: High-order methods based on collocation techniques*, J. Integral Equations, 6 (1984), pp. 187–203.
P. P. B. EGGERMONT, *Collocation as a projection method and superconvergence for Volterra integral equations of the first kind*, in Baker and Miller [22], pp. 131–138.

———, *Collocation for Volterra integral equations of the first kind with iterated kernel*, SIAM J. Numer. Anal., 20 (1983), pp. 1032–1048.

C. J. GLADWIN, *On optimal integration methods for Volterra integral equations of the first kind*, Math. Comp., 39 (1982), pp. 511–518.

E. HAIRER, *Extended Volterra–Runge–Kutta methods*, in Baker and Miller [22], pp. 221–231.

E. HAIRER AND CH. LUBICH, *On the stability of Volterra–Runge–Kutta methods*, SIAM J. Numer. Anal., 21 (1984), pp. 123–135.

P. J. VAN DER HOUWEN AND H. J. J. TE RIELE, *Linear multistep methods for Volterra integral equations of the second kind*, in Baker and Miller [22], pp. 79–94.

Z. JACKIEWICZ, *The numerical solution of Volterra functional differential equations of neutral type*, SIAM J. Numer. Anal., 18 (1981), pp. 615–643.

D. KERSHAW, *Some results for Abel–Volterra integral equations of the second kind*, in Baker and Miller [22], pp. 273–282.

———, *The stability of a numerical method for a second kind Abel equation*, in Baker and Miller [22], pp. 459–461.

CH. LUBICH, *On the stability of linear multistep methods for Volterra equations of the second kind*, in Baker and Miller [22], pp. 233–238.

———, *On the stability of linear multistep methods for Volterra convolution equations*, IMA J. Numer. Anal., 3 (1983), pp. 439–465.

A. MAKROGLOU, *Hybrid methods in the numerical solution of Volterra integro-differential equations*, IMA J. Numer. Anal., 2 (1982), pp. 21–35.

———, *A block-by-block method for the numerical solution of Volterra delay integro-differential equations*, Computing, 30 (1983), pp. 49–62.

P. MARKOWICH AND M. RENARDY, *The numerical solution of parabolic Volterra equations arising in polymer rheology*, SIAM J. Numer. Anal., 20 (1983), pp. 890–908.

———, *Lax–Wendroff methods for hyperbolic history value problems*, SIAM J. Numer. Anal., 21 (1984), pp. 24–51.

S. MCKEE, *A review of linear multistep methods and product integration methods and their convergence for first kind Volterra integral equations*, in Baker and Miller [22], pp. 153–162.

S. MCKEE AND A. STOKES, *Product integration methods for the nonlinear Basset equation*, SIAM J. Numer. Anal., 20 (1983), pp. 143–160.

P. J. TAYLOR, *Applications of results of Vainikko to Volterra integral equations*, in Baker and Miller [22], pp. 185–196.

H. J. J. TE RIELE, *Collocation methods for weakly singular second kind Volterra integral equations with nonsmooth solution*, IMA J. Numer. Anal., 2 (1982), pp. 437–449.

P. H. M. WOLKENFELT, *Reducible quadrature methods for Volterra integral equations*, in Baker and Miller [22], p. 67.

———, *The construction of reducible quadrature rules for Volterra integral and integro-differential equations*, IMA J. Numer. Anal., 2 (1980), pp. 131–152.

———, *On the relation between the repetition factor and the numerical stability of direct quadrature methods for second kind Volterra integral equations*, SIAM J. Numer. Anal., 20 (1983), pp. 1049–1061.

———, *Modified multilag methods for functional equations*, Math. Comp., 40 (1983), pp. 301–316.

Index

Abel's equation:
  generalized, 5
  inversion formulas, 74
  numerical solution, 165
  simple, 4
accumulated error:
  consistency, 104
  starting, 104
Adams-Moulton methods, 182
Aitken's method, 134
A-stability, 111
asymptotic behavior of solutions, 42, 61, 89

Beltyukov methods, 122
block-by-block methods:
  for Abel's equation, 169
  for equations of the first kind, 154
  for equations of the second kind, 114
  for integrodifferential equations, 185
  using product integration, 136
B-splines, 209

causality, 17
characteristic polynomial of stability, 112
comparison theorems, 40, 47, 58
consistency condition:
  for integral equations, 101
  for integrodifferential equations, 183
consistency error:
  accumulated, 104
  local, 101
  order, 101
contraction mapping argument, 32
convergence:
  and stability, 185
  of an approximation method, 100
  order of, 101

conversion:
  differential equation to integral equation, 13
  first kind equation to second kind, 67, 68
  integral equation into differential equation, 7, 78
  integrodifferential equation to integral equation, 5
convolution equation:
  linear, 18, 77
  nonlinear, 77
convolution theorem, 84

Dahlquist theory, 177
degenerate kernel approximation methods, 126
delay differential equation, 189
difference kernel, 37
differential resolvent, 83
differentiation methods for first kind equations, 158
direct methods for equations of the first kind, 143
discretization error, 100

effect of data errors:
  on Abel's equation, 204
  on equations of the first kind, 204
elongation of a filament, 204
error estimates, 103
error expansion, 105, 121
Euler's method, 133, 144
existence of a solution:
  Abel's equation, 73, 75
  equations with unbounded kernels, 48
  integrodifferential equations, 49
  linear first kind equations, 67
  linear second kind equations, 29
  nonlinear first kind equations, 69
  nonlinear second kind equations, 52, 55

INDEX

systems of the second kind, 48

finite rank kernel, 8
first kind equations:
  nonlinear, 69, 160
  numerical solution, 143
  with smooth kernels, 67
Fourier transform, 19
functional equations, 188

Gregory methods:
  for equations of the second kind, 98
  for equations of the first kind, 151

heat conduction equation, 19

ill-posed problems, 162
initial value problems and Volterra equations, 7
instability in numerical computation, 103
integral equations:
  of Volterra type, 3
  of Fredholm type, 3
  singular, 3
  weakly singular, 3
integrodifferential equations:
  reduction to integral equations, 5
  numerical methods, 177
interchanging order of integration, 6
inversion formulas:
  for Fourier transforms, 19
  for Laplace transform, 84
iterated kernels, 36

kernel:
  degenerate, 8
  finite rank, 8
  Lipschitz continuous, 51
  monotonicity properties, 58
  of an integral equation, 4
  unbounded, 4

Laplace transforms, 84
linear multistep methods, 182
Lipschitz condition:
  for equations of the first kind, 69
  for equations of the second kind, 52
  for integrodifferential equations, 177

method of continuation, 32
method of successive approximation, 29
midpoint method:
  for Abel's equation, 166
  for equations of the first kind, 144
Milne-Simpson methods, 182

Newton-Cotes methods:
  for first kind equations, 152
  for second kind equations, 98
numerical stability:
  equations of the first kind, 149
  equations of the second kind, 108, 110
  integrodifferential equations, 186

Pascal program:
  trapezoidal method for second kind equations, 195
  product trapezoidal method for second kind equations, 196
  midpoint method for first kind equations, 197
  product midpoint method for first kind equations, 198
perturbation, effect:
  on first kind equations, 71, 162
  on second kind equations, 45
Picard method, 29
piecewise polynomial approximations, 118, 157
polymer rheology example, 204
population equation, 15
predictor-corrector methods, 125
product integration, 130
product trapezoidal method:
  equations of the second kind, 135
  equations of the first kind, 168
Pouzet methods, 122

radiating source, 23
reactor dynamics equation, 21
renewal:
  equation, 14
  density, 14
repetition factor, 106
resolvent equation:
  difference kernels, 38
  linear problems, 36
  nonlinear problems, 63
resolvent kernel, 35
Richardson's extrapolation, 133
roundoff error:
  in first kind equations, 161
  in second kind equations, 121
Runge-Kutta methods:
  explicit, 122

implicit, 114
of Beltyukov type, 122
of Pouzet type, 122

second kind equations:
  linear, 29
  nonlinear, 51
  numerical solution, 95
  product integration methods, 129
semicirculant matrix, 172
Simpson's methods, 99
starting values, 98
starting error, 104
stability:
  definition, 104, 183
  interval, 111
stereology example, 24
system identification, 18
systems:
  causal, 17
  evolutionary, 3
  feedback, 18
  history-dependent, 13
  linear, 16
  nonlinear, 16
  time-invariant, 17
  with memory, 13
systems of integral equations, 5, 46
systems theory, 15

Tauberian theorems, 91

trapezoidal method:
  Abel's equation, 166
  equations of the second kind,
  equations of the first kind, 144
  integrodifferential equations, 178

uniqueness of a solution:
  Abel's equation, 73, 75
  equations with an unbounded kernel, 48, 62
  integrodifferential equation, 49
  linear first kind equation, 67
  linear second kind equation, 29
  nonlinear first kind equation, 69
  nonlinear second kind equation, 52, 55
  systems of the second kind, 46, 62
unit-impulse function, 16
unit-response function, 17

Volterra equations:
  applications, 13
  classification, 3
  linear, 4
  of first kind, 4
  of second kind, 3
Volterra operators, connection with differential operators
Volterra population equation, 15

weights for numerical integration, 98
well-posed problems, 45